I0015321

Hybrid Cloud Security Patterns

Leverage modern repeatable architecture patterns to secure your workloads on the cloud

Sreekanth Iyer

BIRMINGHAM—MUMBAI

Hybrid Cloud Security Patterns

Group Product Manager: Rahul Nair
Publishing Product Manager: Niranjan Naikwadi
Senior Editor: Athikho Sapuni Rishana
Technical Editor: Nithik Cheruvakodan
Copy Editor: Safis Editing
Project Coordinator: Ashwin Kharwa
Proofreader: Safis Editing
Indexer: Subalakshmi Govindhan
Production Designer: Prashant Ghare
Marketing Coordinator: Nimisha Dua

First published: December 2022
Production reference: 1201022

Published by Packt Publishing Ltd.
Livery Place
35 Livery Street
Birmingham
B3 2PB, UK.

ISBN 978-1-80323-358-1

www.packt.com

To my mother Parvathy and my father Ramakrishna Iyer for their sacrifices and the values they have instilled in me – to stay selfless, work hard, and be thankful.

To my wife Saritha and my sons Varun and Vignesh for their love, support, and inspiration.

– Sreekanth Iyer

Foreword

As enterprises and businesses adopt hybrid cloud to accelerate innovation, cloud security remains an important focus area to both mitigate risk and achieve compliance. Leveraging his hands-on experience in building cloud delivered products, as well as solution engagements with customers to address their challenges, Sreekanth has done a wonderful job in outlining a practical approach to cloud security in this book.

Capturing best practices and repeatable patterns is a great way to bring together the different dimensions of cloud security, with practical solutions that are readily usable. For each of the patterns, his approach to outlining use cases, challenges, solution approaches, along with applicable technologies from the different cloud providers, is commendable. Each chapter provides standalone content, rendering the book a readily referenceable asset which is thus very valuable to cloud security practitioners who can quickly get to their topic of interest.

I have worked closely with Sreekanth for more than a decade, and I can clearly see him bringing his expertise, experience, and passion for sharing his knowledge - all wrapped into this book.

Dr. Nataraj Nagaratnam

IBM Fellow, CTO for Cloud Security at IBM

Contributors

About the author

Sreekanth (Sreek) Iyer is a thought leader in architecture with over 25 years of experience building enterprise solutions across multiple industries. He is currently working as a principal architect with Apptio. Prior to this role, he worked as an executive IT architect at IBM. He has served as a trusted advisor on digital transformation strategies and the journey to the cloud for many enterprise clients. He is an expert in cloud engineering, security, complex integration, and app modernization. He is an IBM Master Inventor with more than 60 patents. He has built strong software engineering teams and made outstanding contributions to creating security reference architectures. When he is not working, he enjoys music and his time with family and friends.

My sincere thanks to Nataraj Nagaratnam and Sridhar Muppidi at IBM for introducing me to the world of security and for their continued guidance and support.

I'm grateful to Marc Fiammante for being my career mentor and inspiration to write this book. My gratitude to Kyle Brown and Bobby Woolf for imparting the knowledge on pattern language.

I'm thankful to Tony Carrato for the careful and detailed technical review of the book that helped significantly improve the quality of the content. I'm very fortunate to have Tony, who has extensive experience and deep expertise in the cloud security domain, as the technical reviewer .

I've benefited from every interaction with my IBM and Apptio colleagues. I've tremendous respect for each of them. This book reflects the knowledge and wisdom gained from engagement and collaboration with my talented colleagues.

Finally, my sincere thanks to the Packt publishing team – Neil, Niranjan, and Sapuni for their patience, support during difficult times, and their constant encouragement to complete this project.

About the reviewer

Tony Carrato is a member of the steering committee of the Security Forum at The Open Group, as well as an invited expert in their Security Forum. He is a member of the planning group for the New Mexico Technology Council's Cybersecurity Peer Group and a part of the Critical Asset Management (for climate resilience) open source project. He is on the board of Telemetry Insight, a New Mexico startup, and a board advisor to the Ortelius open source project focused on microservices and software supply chain security.

He retired from IBM in 2019, with a total of nearly 50 years of technology experience. His major areas of expertise are in technology architecture, including security, enterprise, and solution architecture.

I've known and worked with Sreek for many years. He's truly knowledgeable about security and the cloud and very good at explaining difficult topics in the area of hybrid cloud security. It's been a pleasure and privilege to support this book coming to fruition.

Table of Contents

Part 2: Identity and Access Management Patterns

Part 3: Infrastructure Security Patterns

5

How to Secure Compute Infrastructure   101

6

Implementing Network Isolation, Secure Connectivity,
and Protection   125

Part 4: Data and Application Security Patterns

7

8

Part 5: Cloud Security Posture Management and Zero Trust Architecture

9

Managing the Security Posture for Your Cloud Deployments   193

10

Building Zero Trust Architecture with Hybrid Cloud Security Patterns   205

Preface

Hybrid cloud security is a complex topic and needs different considerations in various security domains. People who are new to the topic can master the subject in no time with a pattern-based approach. *Hybrid Cloud Security Patterns* is a comprehensive introduction to cloud security patterns.

This book discusses security patterns and how to implement them, with specific cloud providers and pointers to tutorials and easy-to-follow prescriptive guidance. It comes complete with pointers to tutorials and guidance on how to secure or implement security patterns on specific clouds – AWS, Azure, GCP, and IBM Cloud.

By the end of this book, you will learn to use the power of patterns to address security for all your cloud deployments.

Who this book is for

This is a guide for cloud solution architects and security focals to securely deploy their applications in the cloud. This provides prescriptive guidance for cloud engineers/DevSecOps professionals who can build security by design for their cloud-native applications. This also provides business users who are considering cloud deployments with the different aspects of security that they need to consider.

What this book covers

Chapter 1, Opportunities and Challenges with Hybrid Multi-cloud Solution, discusses the evolution of cloud, cloud consumption and deployment patterns, challenges, and opportunities.

Chapter 2, Understanding Shared Responsibility Model for Cloud Security, discusses an overall approach to addressing hybrid cloud security.

Chapter 3, Implementing Identity and Access Management for Cloud Users, describes the patterns to implement authentication, access control, and audit for cloud resources.

Chapter 4, Implementing Identity and Access Management for Applications, shows you how to add authentication and access to web and mobile applications deployed in the cloud. This chapter will discuss the pattern to enhance apps with advanced security capabilities.

Chapter 5, How to Secure Compute Infrastructure, shows you how to secure **Virtual Machines (VMs)** and containers. We will discuss patterns to provide isolation to varying degrees and enable both portability and security for VMs and containers.

Chapter 6, Implementing Network Protection, Isolation, and Secure Connectivity, discusses how to secure a cloud network and the architecture patterns and security elements needed to secure the network, including isolation, connectivity, and protection.

Chapter 7, Data Protection Pattern, explores data protection patterns, including protecting data at rest, in transit, and in use. Data at rest protection patterns include how to protect files, objects stored physically in a database, or raw, in data or storage services. You will learn how to use encryption and key management patterns to protect data at rest, and understand the threats related to data in transit and patterns for protecting data in transit. This chapter will discuss the importance of certificates and their use in protecting data in transit. This chapter also discusses how to protect data during processing, as well as services from the cloud that deliver stronger end-to-end data security in the cloud.

Chapter 8, Shift Left Security for DevOps, discusses how to infuse security into a DevOps pipeline. Shifting left security to be incorporated in the early first stages of concept, development, and operations is required to ensure an application runs safely in the cloud. Threat and vulnerability management are critical aspects of security and compliance programs. This chapter discusses patterns to identify vulnerabilities in cloud resources across infrastructure, middleware, and applications and how to remediate them. Configuration management is another important topic that covers how to manage and control configurations for cloud resources to enable security and facilitate the management of risk.

Chapter 9, Manage Security Posture for Your Cloud Deployments, delves into **Cloud Security Posture Management (CSPM)**, which helps to proactively monitor, track, and react to security violations. This chapter provides information on how to build end-to-end visibility and integration of security processes and tooling throughout an organization to get a security posture for cloud applications. A security and compliance posture provides a method to see controls in place against policies and their effectiveness. This chapter discusses how to prepare an enterprise to respond to large volumes of alerts and events related to cloud security. Given the use of multiple tools and a shortage of staff, enterprises need to adopt security orchestration, automation, and response to improve their effectiveness against security events.

Chapter 10, Building Zero Trust Architecture with Hybrid Cloud Security Patterns, discusses reference architectures and patterns to implement the zero trust model. The principles for zero trust are also discussed in detail. This chapter explores the use cases requiring the zero trust model and how to leverage hybrid cloud security patterns to protect critical data using zero trust security practices.

To get the most out of this book

The book assumes you have basic knowledge of the cloud and its advantages. Knowledge of the different types of cloud, their deployment, and consumption models is a pre-requisite.

Examples of security solutions natively provided by cloud service providers covered in this book	Other security solution providers referred to
Amazon Web Services (AWS)	Aqua Security
Microsoft Azure	Palo Alto Networks
Google Cloud Platform (GCP)	Check Point Software Technologies
IBM Cloud	McAfee, Red Hat, CyberArk, VMware, Intel, Cisco, Tokenex, CloudCheckr, SonarQube, ColorTokens, Zscaler, ServiceNow

The GitHub repository provides links that provide details on how to implement the patterns discussed in each chapter. Refer to Git pages and follow the links on the tutorials and examples from the security services and solution providers listed above. If you are using the digital version of this book, we advise you to type the code yourself or access the reference links to examples code from the book's GitHub repository (a link is available in the next section). Doing so will help you avoid any potential errors related to the copying and pasting of code.

Download the example code files

You can download the example code files for this book from GitHub at `https://github.com/PacktPublishing/Hybrid-Cloud-Security-Patterns`. If there's an update to the code, it will be updated in the GitHub repository.

We also have other code bundles from our rich catalog of books and videos available at `https://github.com/PacktPublishing/`. Check them out!

Download the color images

We also provide a PDF file that has color images of the screenshots and diagrams used in this book. You can download it here: `https://packt.link/cbJMK`.

Conventions used

There are a number of text conventions used throughout this book.

`Code in text`: Indicates code words in text, database table names, folder names, filenames, file extensions, pathnames, dummy URLs, user input, and Twitter handles. Here is an example: "Mount the downloaded `WebStorm-10*.dmg` disk image file as another disk in your system."

Bold: Indicates a new term, an important word, or words that you see on screen. For instance, words in menus or dialog boxes appear in **bold**. Here is an example: "Select **System info** from the **Administration** panel."

> **Tips or Important Notes**
> Appear like this.

Get in touch

Feedback from our readers is always welcome.

General feedback: If you have questions about any aspect of this book, email us at `customercare@packtpub.com` and mention the book title in the subject of your message.

Errata: Although we have taken every care to ensure the accuracy of our content, mistakes do happen. If you have found a mistake in this book, we would be grateful if you would report this to us. Please visit `www.packtpub.com/support/errata` and fill in the form.

Piracy: If you come across any illegal copies of our works in any form on the internet, we would be grateful if you would provide us with the location address or website name. Please contact us at `copyright@packt.com` with a link to the material.

If you are interested in becoming an author: If there is a topic that you have expertise in and you are interested in either writing or contributing to a book, please visit `authors.packtpub.com`.

Share Your Thoughts

Once you've read *Hybrid Cloud Security Patterns*, we'd love to hear your thoughts! Scan the QR code below to go straight to the Amazon review page for this book and share your feedback.

https://packt.link/r/1803233583

Your review is important to us and the tech community and will help us make sure we're delivering excellent quality content.

Download a free PDF copy of this book

Thanks for purchasing this book!

Do you like to read on the go but are unable to carry your print books everywhere? Is your eBook purchase not compatible with the device of your choice?

Don't worry, now with every Packt book you get a DRM-free PDF version of that book at no cost.

Read anywhere, any place, on any device. Search, copy, and paste code from your favorite technical books directly into your application.

The perks don't stop there, you can get exclusive access to discounts, newsletters, and great free content in your inbox daily

Follow these simple steps to get the benefits:

1. Scan the QR code or visit the link below

https://packt.link/free-ebook/9781803233581

2. Submit your proof of purchase

3. That's it! We'll send your free PDF and other benefits to your email directly

Part 1: Introduction to Cloud Security

Security is the primary concern for enterprises adopting hybrid IT and multi-cloud technologies as they pursue application modernization. By taking a strategic approach to security, businesses can infuse security into various stages of their journey to the cloud. This part will discuss how enterprises are adopting hybrid cloud and the challenges with regard to securing their transition to the cloud.

This part comprises the following chapters:

- *Chapter 1, Opportunities and Challenges with Hybrid Multi-cloud Solution*
- *Chapter 2, Understanding Shared Responsibility Model for Cloud Security*

1
Opportunities and Challenges with Hybrid Multi-cloud Solutions

Businesses are rapidly transforming to the digital era. Companies are reinventing processes and cultures to deliver enhanced experience to their customers using digital technologies. This drives the need to build new capabilities and modernize existing applications using the latest technology more quickly. Enterprises are trying to stay ahead of the competition. Being late to market can mean missed opportunities, lost revenue, or, even worse, going out of business. Companies who have been agile and successful are leveraging cloud at the heart of this digital transformation. Furthermore, they are taking a hybrid multi-cloud strategy and approach consisting of on-premises, private, and public clouds to drive better efficiency, performance, and cost optimization. For a business rapidly transforming into a digital enterprise that relies on a hybrid multi-cloud environment to do so, the security threats and attack surface become greater. It is critical to stay ahead of threats, protect valuable data and resources, and achieve regulatory compliance. This chapter discusses digitization trends, the hybrid cloud strategy adopted by enterprises, and the related security challenges.

In this chapter, we're going to cover the following topics:

- The evolution of the cloud
- The digitization trends that drive opportunities and challenges for hybrid cloud solutions
- Security in the digital hybrid multi-cloud era

The evolution of the cloud

Driven by trends in the consumer internet, cloud computing has become the preferred way to consume and deliver IT solutions and services. Before we dive deeper into cloud security, it is important to understand some basic aspects of the cloud, the emerging trends in cloud solutions, culture, technologies, and modern development and delivery models.

Defining cloud computing

Let's start by understanding and defining the term **cloud computing** in detail. It comprises two words – cloud and computing. So, simply put, it is computing that you can offer on the cloud. What exactly is the cloud referred to here? IT architects used the cloud symbol to represent the internet or the network in their drawings. The term **cloud** has evolved as a metaphor for the internet. Computing could be any goal-oriented activity requiring or benefiting from the usage of IT, which includes hardware and software systems used for a wide range of purposes – collecting, storing, processing, and analyzing various kinds of information. Cloud computing has evolved over time from utility computing to what it is today, enabled by virtualization, automation, and service orientation.

The following diagram defines the key elements of cloud computing:

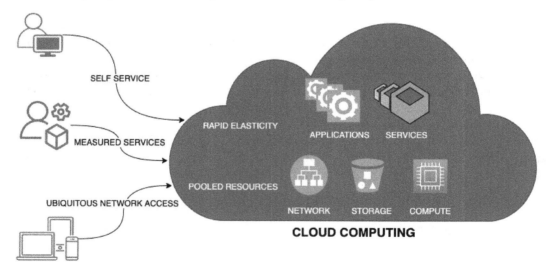

Figure 1.1 – Cloud computing

There are several definitions that you can find on the web for cloud computing. **National Institute of Standards and Technology (NIST)** has promoted the effective and secure use of cloud computing technology within government and industry by providing technical guidance and promoting standards. According to NIST, cloud computing is a pay-per-use model of enabling available, convenient, and on-demand network access to a shared pool of configurable computing resources (for example, networks, servers, storage, applications, and services) that can be rapidly provisioned and released with minimal management effort or service provider interaction. In general, most of the companies have agreed on certain general characteristics or essentials that NIST has pioneered that qualify any internet-based computing to be referred to as a cloud. They are the following:

- **On-demand self-service**: Cloud computing provides a catalog through which a consumer can request to provision any kind of service – computing involving a server, network, and storage or a middleware service such as a database or a software service such as email. This catalog provides self-service without requiring manual intervention on the part of the service provider.

- **Ubiquitous network access**: The key premise of cloud computing is that all the services and capabilities provided are accessible through the network. This can be the internet in the case of a public cloud or the intranet in the case of a private cloud. The resources on a cloud can be accessed through a variety of devices such as computers, mobile phones, and IoT devices over the network through multiple protocols.

- **Location-independent resource pooling**: A cloud's business value comes from the economy of scale that is achieved by resource pooling. The provider pools the available computing resources and makes them dynamically available to clients based on demand. Physical resources including compute, network, and storage are pooled and leveraging virtualization assigned to clients in a multi-tenant model. In certain cases, consumers may not even know the exact location of the provided resources.

- **Rapid elasticity**: The cloud provides a means to rapidly scale up or scale down based on the demand. For the consumer, this is a very valuable business advantage of cloud solutions, as it requires them to only invest in resources when they need to. For instance, cloud consumers can start small with addressing requirements for one region or country and then scale their operations across the globe. Modern cloud technologies offer running applications and managing data without having to worry about infrastructure. Technologies such as serverless computing provide rapid elasticity and scale at a lower cost.

- **Pay per use**: Each cloud service is monitored, metered, and facilitates chargeback. This allows providers to promote their subscription plans and consumers to choose a billing model that is optimal for their resource usage. One example is a time-based pricing model – a per hour, per minute, or per second basis for resources such as servers. A tiered pricing model provides consumers to choose a plan from a set of price points that map to their volume or period of consumption – such as for storage, network bandwidth, or data used. Certain other services such as authentication or validation services can be consumed from the cloud with a plan that is based on active user accounts per month. The chargeback to specific departments inside the organization is now also possible with an accounting model supported by the providers and the ability to tag cloud resources to specific departments.

Cloud personas

There are several actors typically involved in building and operating a cloud solution. Their roles and responsibilities and their relationships with other actors vary based on the industry:

- **Business owners**: This actor's responsibilities are to make appropriate cloud investment decisions. This section is more focused on the innovation and agility that the cloud can provide for their business. Once an organization has started with cloud solutions, then there are some typical actors that are involved in the day-to-day operational consumption and provision of cloud services.

  Cloud personas and their roles are shown in the following diagram and described in the section that follows:

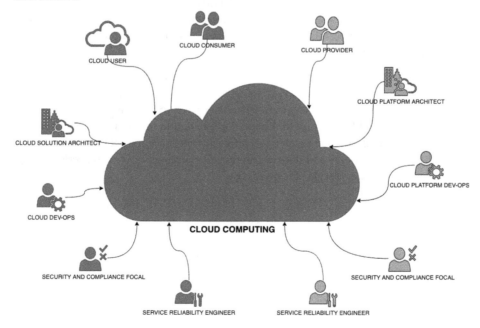

Figure 1.2 – Cloud personas

- **Cloud service consumer**: The enterprise or end user who subscribes and uses the cloud-based application or service.

- **Cloud service provider**: The organization that defines, hosts, and delivers cloud computing services to its consumers.

- **Cloud service creator or developer**: The organization or developer who creates and publishes the cloud service on a catalog for consumption.

Out of all the roles across all these organizations, the key roles from an implementation and operation perspective are the following:

- **Cloud administrator** who can perform the following tasks:

  - Setting up the cloud account(s) for the organization

  - Defining the users, teams, and their associated roles

  - Allocating or defining the quota for projects and users with the associated charges

  - Approving or denying requests for provisioning or de-provisioning cloud resources

  - Monitoring consumption by project

- **Cloud user**: Accesses or uses the cloud deployed applications, services, or provisioned resources (for example, the application, storage, or servers available to them).

There are variations within these two roles depending on the cloud provider and consumer organization design. There is more rationalization of these traditional roles in the modern context. These roles include the following:

- **Cloud solution architect**: The person with the knowledge and skills on how to design applications that can effectively leverage cloud capabilities. They understand specific cloud environments, such as AWS, Azure, IBM, and Google, and leverage their services and technologies to build highly scalable, performant, and available applications.

- **Cloud DevOps engineer**: A cloud user who is primarily responsible for developing the application component or service. The Dev-Ops engineer is also responsible for building the pipeline to deploy, monitor, and operate the service. DevOps speeds up software development and delivery, bringing close collaboration with engineering and operations teams.

- **Service Reliability Engineer** (**SRE**): Primarily responsible for improving the reliability of services through collaboration with development, proactive monitoring, and optimization of redundancies in operations. SRE is an integral part of modern cloud development teams who are involved in proactive testing, observability, service reliability, and speed.

- **Security and compliance focal**: Core members of the cloud teams who ensure the services are designed, developed, and deployed securely on the cloud. Ensuring services meet regulatory and security compliance requirements is the responsibility of the security and compliance focal. These resources define security policies and procedures, execute audit checks and governance related to backup, and restore automation for security and compliance tasks.

Cloud deployment models

Driven by trends in the consumer internet, cloud computing has become the preferred way to consume and deliver IT services. The cloud supports multiple deployment models based on the given requirements. The capabilities delivered by cloud are accessible via a cloud catalog and categorized based on the IT service delivered. These integrated services or layers of IT-as-a-Service are often referred to as **cloud deployment models**. The details of each of the cloud deployment models are shown in the following diagram:

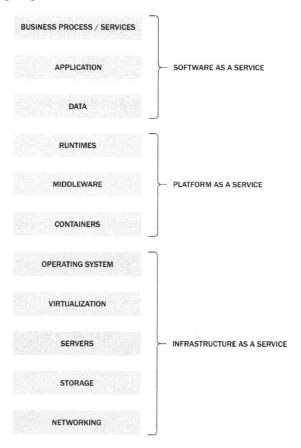

Figure 1.3 – Cloud deployment models

The different deployment models are as follows:

- **Infrastructure as a Service (IaaS)**: In this service delivery model, IT infrastructure is delivered over the network to consumers. This includes the compute (servers), network, storage, and any other data center resources. IaaS provides the ability to rapidly scale up or scale down infrastructure resources. IaaS consumers can concentrate on deploying and running their software, services, or applications without having to worry about managing or controlling the underlying resources.

- **Platform as a Service (PaaS)**: Provides a platform for consumers to develop and deploy their applications. While IaaS provides the infrastructure resources, PaaS provides the programming languages, tools, and platforms to develop and deploy applications. Consumers have the ability the to control deployed applications and operating systems and environments.

- **Software as a Service (SaaS)**: The cloud deployment model where application and services are made available to clients. In this scenario, customers can use a service without having to worry about the development, deployment, or management of these applications. In the SaaS model, the provider takes care of making the applications available to multiple clients. End users need not install or manage any software on their side and can access the applications through their devices of choice. Popular services or applications provided in the SaaS model are e-mail, ERP, and CRM.

- **Business Process as a Service (BPaaS)**: An emerging model on top of SaaS where customers can consume business processes such as accounting and payroll, or HR processes such as travel and expense management as a service. These business services are accessed via the internet and support multiple subscription plans as advertised by the provider. The consumer can choose from these plans and subscribe to the services based on their requirements.

Cloud delivery models

The support for different delivery models is the critical success factor of the cloud for business. The flexible cloud delivery models or cloud types are shown in the following diagram:

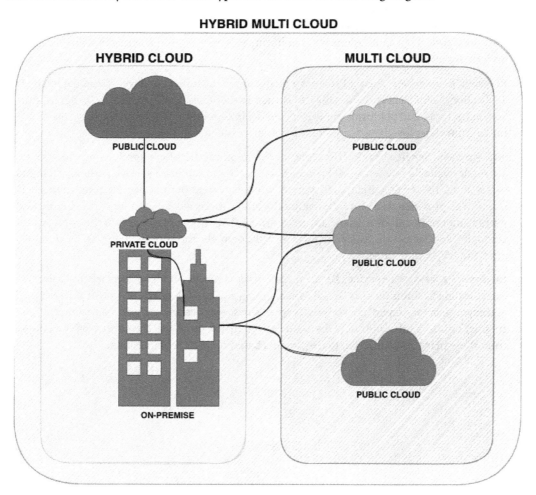

Figure 1.4 – Cloud types (delivery models)

We shall see the various types in detail:

- **Private cloud**: Refers to resource pooling and sharing IT capabilities within the enterprise or behind a firewall. These are often managed privately and run by the enterprise itself and made available to the users on their intranet. A private cloud provides more flexibility to the enterprise in terms of the customization of services. At the same time, a private cloud also drives internal efficiency, standardization, and best practices. Since the resources and management are mostly local or dedicated, private cloud provides tighter control and visibility.

- **Public cloud**: Refers to a standardized set of business, application, or IT services provided *as a service* over the internet. In this model, the service provider owns and manages the service and access is by subscription. Multitenancy is a key characteristic of public cloud services that enable economies of scale. The flexible price per use basis is applicable and greater discounts apply to a committed higher usage.

- **Hybrid cloud**: Combines the characteristics and delivery models of both public and private clouds. The hybrid cloud as a solution combines the best of all worlds – on-premises, private clouds, and multiple public cloud services. In a hybrid cloud model, a solution can have components running on-prem on a private cloud or enterprise infrastructure connecting to services running on a public cloud. A hybrid cloud strategy is preferred by businesses, as it provides greater flexibility and resiliency for scaling workloads based on demand at reduced cost.

- **Multi-cloud**: Refers to leveraging services provided by more than one cloud – refers to the use of private and public services and their integration. A business may have multiple services across IaaS, PaaS, and SaaS provided by multiple vendors. A multi-cloud approach consists of a mix of major public cloud providers or hyperscalars, namely **Amazon Web Services (AWS)**, **Google Cloud Platform (GCP)**, **Microsoft (Azure)**, and **IBM**.

- **Hybrid multi-cloud**: If the multi-cloud deployment includes a private cloud or an on-premise data center as well, then the cloud deployment can actually be considered a hybrid multi-cloud. We also see several variations of adoption of these cloud delivery and consumption models.

From cloud to hybrid multi-cloud

More cloud service types are emerging and guiding the development of the IT industry. These multiple delivery models can co-exist and integrate with traditional IT systems. The cloud type and delivery model selection depend on the workload and the intended benefits.

The key intended benefits from cloud are as follows:

- **Speed**: Capability to provision on demand and elastically scale computing resources (infrastructure, platforms, software, and business services). This is enabled through automated secure and managed provisioning process. Most cloud computing services are provided through self-service catalogs on demand. A big set of computing resources and environments can be automatically provisioned in minutes instead of having to wait for hours and days. The delivery of services more rapidly is enabled with automation and less human intervention. With proper automation, this ensures fewer errors and fulfillment of requested qualities of service or **Service Level Agreements (SLAs)**.

- **Cost**: Enterprises don't have to invest in buying hardware and software for their data centers, as well as incurring the cost of managing these resources. Depending on the delivery and consumption model, the cost and security of the cloud are defined through a shared responsibility matrix that's documented and reviewed regularly. The cloud provides a way to cut down on the enterprise **capital expenses (Capex)** on racks, servers, cooling, electricity, and the IT service professionals for managing the infrastructure. The cloud provides a more efficient pricing model and lowers both capital and operational expenditure.

- **Flexibility**: Businesses need to adjust the IT resources based on the market demands. They need to balance performance, security, availability, and scale based on the business requirements. The cloud provides a seamless and efficient way to manage availability, resilience, and security with flexibility to move workloads across on-premise, private, and public infrastructures and services.

- **Resiliency**: Improved risk management through improved business resiliency. Improved time to market and acceleration of innovation projects. Cloud computing makes data backup, disaster recovery, and business continuity seamless and inexpensive with multiple availability zones on a cloud provider's network.

- **Efficiency and global scale**: The benefits of cloud computing services include the ability to scale elastically. That means rapidly expanding to new geographies with the right amount of IT resources. The cloud not only optimizes the IT resources but also frees up time for skilled resources to focus on innovative and future-looking projects. The cloud helps significantly improve energy efficiency through sharing and the optimal usage of resources. The cloud infrastructure and services are upgraded to the latest ones at a faster pace to provide fast and efficient computing hardware and services. This offers several benefits over traditional data centers, including reduced network latency for applications with multiple availability zones and greater economies of scale.

Most enterprises start with something under their control to optimize what is behind their firewalls. So, the initial interest was tremendously geared toward private clouds – in both large enterprises and the mid-market. There was great interest initially in public cloud services for infrastructure services especially. Businesses have become comfortable moving workloads externally with domain applications available on the public cloud. This has resulted in a proliferation of hybrid clouds with the need for businesses to integrate their private environments with public cloud services.

Digitization trends

Enterprises are seeking to get a deeper understanding of their data and provide differentiated, personalized experiences for their employees, customers, and partners. This requires modern applications to be created that are more responsive and can be used by clients across different types of devices. This also requires collecting a lot more data and applying artificial intelligence and machine learning to create personalized insights. This experience must be highly scalable, available, and made available for large set of users. This means it has to be built and managed on hybrid multi-cloud platforms leveraging an automated DevOps pipeline. We will discuss the impact of this digital transformation across architecture, application, data, integration, management, automation, development, and operations. Security and compliance are important cross-cutting concerns that needs to be addressed for each of these areas as part of this transformation.

Application modernization

The key opportunities and challenges with application modernization in the context of hybrid cloud are discussed in the following diagram:

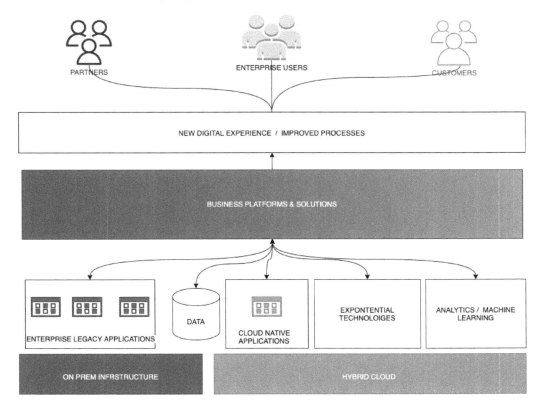

Figure 1.5 – Application modernization

The key trends in application modernization and migration to the cloud are listed as follows:

- **Cloud-native applications**: Companies are building new capabilities as cloud-native applications that are based on microservices. Taking a cloud-native approach provides high availability. Applications are able to launch to the market quickly and scale efficiently by taking a cloud-native approach. Microservices have emerged as the preferred architectural style for building cloud-native applications. This structures an application as a collection of services that are aligned to business requirements, loosely coupled, independently deployable, scalable, and highly available.

- **Modernize and migrate existing applications to the cloud**: Companies are looking at various options to modernize their legacy applications and make it ready for the digital era. They are considering the following:

  - Lifting and shifting to the cloud, where the application and associated data is moved directly to cloud without any changes or redesigning.

  - Re-platforming to a virtualized platform.

  - Exposing application data through APIs.

  - Containerizing the existing application.

  - Refactoring or completely rewriting into new microservices.

  - Strangling the monolith and moving to microservices over time, where a large legacy application is restrained from adding new features and incrementally transformed to new services based on cloud-native principles.

- **Built-in intelligence**: Modern apps include cognitive capabilities and participate in automated workflows. We are seeing this in all the domains across banking, telecom, insurance, retail, HR, and government. Bots or applications answer human queries and provide a better experience. This positive technology also is playing a key soldier role in the healthcare sector to help mankind fight the pandemic.

- **Modernizing batch functions in the cloud**: This is another area that is growing rapidly. Some capabilities such as serverless computing and code engines help to build great applications in shorter time frames. These concepts will be discussed in detail in a later chapter. Essentially, these capabilities or application styles help offload otherwise long-running and resource-hungry tasks to asynchronous jobs. Thus, the cloud provides optimized scale and cost efficiency. Batch jobs that take lot of runtime and processing power or cost in costly legacy systems may be done at a much lower cost with these computing styles.

- **Runtimes**: As part of their modernization strategies, enterprises are evaluating how to minimize the cost of ownership and operations of the apps as well. To this end, some of the traditional applications are getting rearchitected to leverage cloud-native capabilities such as cloud functions or code engines. Another big set of applications are those that needs to be sunset or retired because the compilers or runtimes are going out of service. There is a demand for newer runtimes and faster virtual machines with lighter footprints in terms of resource usage.

- **Tools**: 80% of the core applications are still on legacy platforms. For businesses to move them to the cloud requires greater automation. Many intelligent tools for the discovery and extraction of logic and business rules that are more domain- or industry-focused exist.

Data modernization and the emergence of data fabric

We are at a tipping point in history where technology is transforming the way that business gets done. Businesses are intelligently leveraging analytics, artificial intelligence, and machine learning. How businesses collect, organize, and analyze data and infuse **artificial intelligence** (**AI**) will be the key to this journey being successful.

There are several data transformation use case scenarios for a hybrid multi-cloud architecture. Many companies have to modernize monoliths to cloud-native with distributed data management techniques to deliver digital personalized experiences. This involves transforming legacy architecture characterized by data monoliths, data silos, the tight coupling of data, high **Total Cost of Ownership** (**TCO**), and low speeds into new technologies identified by increased data velocity, variety, and veracity. Data governance is another key use case where enterprises need visibility and control over the data spread across hybrid multi-cloud solutions, data in motion, and data permeating enterprise boundaries (for example, the blockchain). Enterprises need to better efficiency and resiliency, with improved security for their data middleware itself and the workloads.

The greatest trend is related to delivering personal and empathetic customer service and experiences anchored by individual preferences. This is nothing new but how we enable this in a hybrid cloud is important. Customers have their **System of Records** (**SoRs**) locked in filesystems and traditional databases. There is no easy way to introduce new digital products that leverage these legacy data sources. What is required is to have a strategy to unlock the data in the legacy system to participate in these digital plays. To do this, enterprises take multiple approaches, which are listed as follows:

- One is to leave the SoRs or the **Source of Truth** (**SoT**), wherever it is, and add the ability to access the data through APIs in order to participate in digital transactions.

- A second option is to move the core itself to the cloud, getting rid of the constraints to introduce new products more quickly with data on the cloud in the process. In this model, a single legacy database could end up as multiple databases on the cloud. This will remove the dependency on data for the new development teams, thus speeding up the delivery of new digital capabilities.

Another key observation is that the line between transactional and analytical worlds is becoming thinner. Traditionally, enterprises handled their transactional data and analytics separately. While one side was looking into how to scale rapids to support transactions from multiple users and devices, on the other side, analytics systems were looking to provide insights that could drive personalization further. Insights and AI help optimize production processes and act to balance quality, cost, and throughput. This use case also involves using artificial intelligence services to transform the customer experience and guarantee satisfaction at every touchpoint. Within the industrial, manufacturing, travel and transportation, and process sectors, it is all about getting greater insight from the data that's being produced. This data is used to drive immediate real-time actions. So, we see operation analytics (asset management), smarter processes (airport or port management), information from edge deployments, security monitoring driving new insights, and the insights, in turn, driving new types of business.

The emergence of data fabric

Businesses are looking to optimize their investment in analytics to drive real-time results to deliver personalized experiences with improved reliability and drive optimization in operations and response. This is their objective – to raise maturity on the journey to becoming a data-driven organization. Companies invest in industry-based analytics platforms that will drive this agenda related to cognitive capabilities to modernize the ecosystem and enable higher value business functions.

Given the intermingling of transactional data and analytics data, something called the data fabric emerges. It simply means companies don't need to worry about where the data is, how it is stored, and what format it is available in. The data fabric does the hard work of making it available to the systems at the time they need it and in the form required.

The concept and value of a data fabric bridging the transactional and analytics world is detailed in the following diagram:

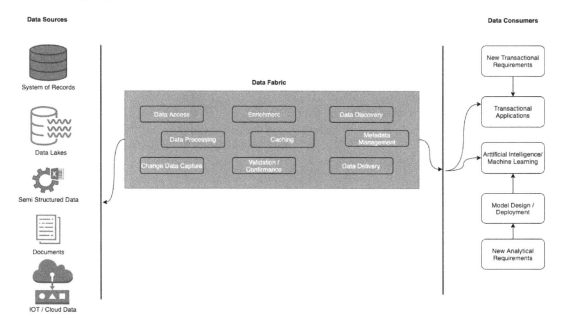

Figure 1.6 – The data fabric

Cloud, data hub, data fabric, and data quality initiatives will be intertwined as data and analytics leaders strive for greater operational efficiency across their data management landscape. Data fabrics remain an emerging key area of interest, enabled by active metadata.

Various architecture aspects of this fabric leverage hybrid multi-cloud containerized capabilities to accelerate the integration of AI infusion aspects with transaction systems.

Lowering the cost of data acquisition and management is another key area that enterprises are dealing with. They are required to manage heterogenous transactional and analytical data environments. So, they need help with decision data and analytics workload dispositions to the cloud. Cost optimization through adopting new cloud-based database technologies is accelerating the move from on-premises databases to cloud-based database technologies. A hybrid cloud strategy providing efficient solutions for how data is stored, managed, and accessed is a prerequisite to these activities.

Transactional and master data management within an organization changes from application to application in terms of its interpretation and granularity. This is another evolving theme of central versus distributed management of data. The fast-adopted model is to synchronize access to this data across the organization orchestrated through a data fabric or DataOps.

Integration, coexistence, and interoperability

The majority of enterprise clients rely on a huge integration landscape. The traditional enterprise integration landscape is still the core of many businesses powering their existing **Business to Business (B2B)**, **Business to Consumer (B2C)**, and **Business to Employee (B2E)** integrations. Enterprises need the agility of their integration layer to align with the parallel advances occurring in their application delivery.

When creating new systems of engagement, cloud engineering teams are looking for new ways to integrate with systems of records. They want to rearchitect, refactor, or reconstruct this middle layer. The demands on this middle layer are increasing as end users demand more responsive apps. Squads building new applications in the form of APIs and microservices like to be in complete control end-to-end, including at the integration layer.

The following diagram shows how the integration layer is modernized to support the needs of a hybrid cloud environment covering coexistence and interoperability aspects as well:

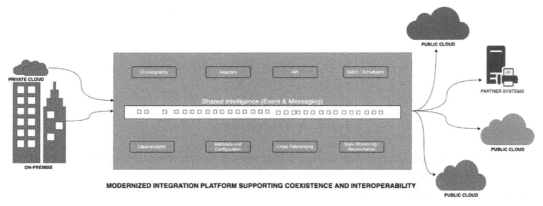

Figure 1.7 – Integration modernization, coexistence, and interoperability

Enterprises are adopting newer hybrid cloud architecture patterns to reach a higher level of maturity with optimized and agile integration. They are also looking for better ways to deploy and manage integrations. There is a need for an optimized CI/CD pipeline for integrations and the capability to roll out increments. Another requirement is controlling the roll-out of an application and integrating code for different target environments. Thus, we see tools for GitOps being extended to the integration layer as well.

Event hubs and intelligent workflows

Clients are looking to tap into interesting business events that occur within their enterprise and an efficient way to respond or react in real-time. The deployment of an event hub and event-based integrations are becoming the new norm. This also is the key enabler for building some of the intelligent workflows for a cognitive enterprise.

This integration model plays an important role in the hybrid cloud world, bringing together otherwise siloed and static processes. This also enables enterprises to go beyond their boundaries to create a platform for customers, partners, and vendors to collaborate and drive actions derived from real-time data. The applications (event consumers and producers) could be within or outside the enterprise. This space is set for exponential growth in the near future.

Coexistence and interoperability

Many enterprises have a clear understanding of their current landscape and the vision for their target multi-cloud architecture. However, they lack the knowledge of what their next logical step to reach the target is. The critical gap that needs to be addressed by their modernization strategy is related to technical and cultural coexistence and interoperability.

In an incremental modernization journey that is continuously iterative, coexistence and interoperability is a natural byproduct. Coexistence architecture isolates the modernized state from the current state. This layer also hides the complexities and gaps, as well as minimizes any changes to current state processing. Coexistence provides a simple, predictable experience and outcome for an organization to transition from old to new, while interoperability is the ability to execute the same functionality seamlessly in the legacy and the modernized side.

Businesses prioritize their investments and focus on acquiring new capabilities and transforming existing capabilities. Coexistence and interoperability help clients to focus on business transformation with logical steps towards realizing their target architecture. With many moving parts, as well as each of them being transformed and deployed on multicloud environments, this is an even more challenging problem to solve. This layer also needs to be carefully crafted and defined for successful multi-cloud adoption.

DevOps

DevOps is a core element of developing and deploying to the cloud. It combines the development and operation of both technological and cultural tools and practices with a perspective to drive the efficiency and speed of delivering cloud applications. It is important to understand the trends in the space and the impact of the adoption of the hybrid cloud.

Optimization of operations

Large enterprises want to ensure uninterrupted business operations with 99.99% application availability. There are traditional huge monolithic applications that can benefit from re-platforming and re-architecturing to the hybrid cloud. However, the goal is not about transformation in itself but about ensuring the cost of operations is minimized. For the FTEs who work on keeping the application alive, businesses need a better way to reduce the effort involved in releases and change management.

Over the years, these enterprises have created solid engineering teams with tons of scripts and automation to build, deploy, and maintain their existing applications. Some of these may be reused, but there is still more automation to be done to reduce manual work and the associated processes.

Leveraging observability for a better customer experience

Over the years, there has been a lot of software or application intelligence gathered by the build and deploy systems. Enterprises are looking to see how to leverage this information to provide a better experience to their clients. This is where observability comes in. Business needs to tap into this telemetry data to determine the aspects of the application that are heavily accessed following client consumption patterns. This also provides problem areas in the application that need to be addressed. Telemetry data-driven integrated operations to monitor across application environments are a requirement in this space. When observability is combined with the efforts of SREs, we have a beautiful way to create highly available solution components on the cloud that can provide a better user experience.

Several enterprises already use AI in this space to manage and understand immense volumes of data better. This includes understanding how the IT events are related and learn system behaviors. Predictive insights into this observability data with AIOps, where artificial intelligence is leveraged for operations and addressing end user engagement with ChatOps or bot-assisted self-service, is a major trend.

Automation, automation, automation

From the provisioning of new sandbox environments through infrastructure as code, development, and testing, automation is the key to do more with fewer developers. As companies build their CI/CD pipeline, automation drives the consistent process to build, run, test, merge, and deploy to multiple target environments. Automation brings QA and developer teams together to move from build, integration, and delivery to deployment. However much automation is done, there will still be a lot of areas in which to improve for enterprises and automation will always be a requirement for this.

Building pipeline of pipelines for hybrid multi-cloud

Continuous planning, **continuous integration** (**CI**), and **continuous delivery** (**CD**) pipeline, as shown in the following diagram, are critical components required to operate software delivery in a hybrid cloud:

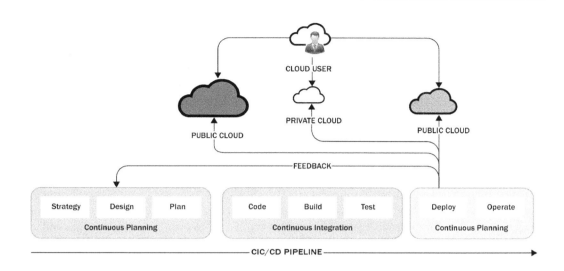

Figure 1.8 – A pipeline of pipelines

Companies need more than one CI/CD pipeline to build and deploy to multiple environments. These DevOps pipelines need to cater to building and deploying to existing and multi-cloud environments. Few are thinking of combining pipelines that can achieve a composite model for hybrid multi-cloud deployment. This is true for applications that have huge middleware and messaging components, which are also going through transformation. Bringing legacy systems into the existing pipeline with common tools and other modern tools is key to enabling application modernization. A robust coexistence and interoperability design sets the direction for an enterprise-wide pipeline. The integration of the pipeline with the service management toolset is another important consideration. Enterprises have to automize with the right set of tools that can also be customized for the target clouds based on the requirements.

Security for the digital hybrid multi-cloud era

Security is a cross-cutting concern across the hybrid enterprise. In this section, we will discuss the challenges and opportunities for hybrid cloud security in the context of digitization trends. This will cover security aspects related to infrastructure, application, data, integration, secure engineering, and operations.

App modernization and security

Businesses must consider security from the beginning when building applications on cloud. They are also looking to take advantage of application modernization as an opportunity to better their overall security posture. threat modeling and secure engineering were always part of traditional application development practices. These practices merit even more importance in a hybrid cloud world, as the attack surfaces and threats are bigger. Clients adopting cloud-native and exponential technologies such as containers, blockchains, and AI need protection and look to secure these technologies as part of their modernization.

There is a need to catch advanced threats and be in a position to respond to them – you need to shift left with threat management in the agile DevOps pipeline. Traditionally, threat management and responding to security incidents are extended components of **Security Information and Event Management (SIEM)**. In the modernization context, we will need to plan for all the ways to combat the different threats during the application development phase itself.

Data security

This important aspect will include how to manage security, protection, and governance of the data on cloud. The data security domain includes controls mapped to security threat landscapes and business contexts – encryption and key management, **Identity and Access Management (IAM)**, vulnerability management, and data activity monitoring.

Another emerging aspect to protect sensitive data in all formats – in transit, at rest, and even in use – is becoming a necessity. This will ensure data in memory is also secure. Use cases such as cryptocurrency and digital assets drive this requirement for performing computation in a hardware-based trusted execution environment. Use cases such as the mining of cryptocurrency shifting the focus from data to computational issues drive the need for tighter security for *data in use*.

Security for integration, coexistence, and interoperability

Integration and interoperability are key enablers for enterprises to build business processes and enable data movement across internal, partner, and supplier systems. In a hybrid multi-cloud environment, integration cuts across applications deployed on-premises, applications in the same cloud, and applications deployed on different clouds, as well as SaaS solutions. Within integration modernization trends, companies are also building the shared intelligence backbone enabled through event streams. The secure transformation of this integration landscape is an essential part of enterprise's digital journey. This will involve securing the different integration patterns leveraged for the purpose – such as API enablement of legacy systems and event-driven architecture, as well as the coexistence and interoperability components.

Shift left security – from DevOps to DevSecOps

With a move from just a few releases a year to weekly feature releases, security can no longer be ensured manually. Security needs to be part of the DevOps pipeline and be automated. There are plenty of security tools out there from various vendors than can integrate with the pipeline. The key things to be addressed in a DevSecOps pipeline include security tools that address the following.

Securing cloud-native development and operations

Securing the DevOps pipeline involves catching security errors early in the cycle and addressing the vulnerabilities of deployable artifacts, as well as performing configuration checks. These aspects are discussed in the following sections.

Helping developers address issues in the code early

A proper **Integrated Development Environment** (**IDE**) should come with source code analysis and code coverage tools that analyze the source code to find security flaws. The usage of security testing tools for identifying potential vulnerabilities (such as OWASP) is a critical element of DevSecOps. There are several open source and vendor tools to that integrate with your pipeline that can secure your application before you deploy to production and ensure they are vulnerability free.

> Information
>
> The **Open Web Application Security Project®** (**OWASP**) is a non-profit foundation that works to improve the security of software. *The OWASP Top 10* is a book or referential document outlining the 10 most critical security concerns for web application security.

Securing deployable artefacts

Artifacts such as containers and third-party libraries need to be scanned for vulnerabilities. The cloud service providers and repository engines typically provide built-in security scanning for these artifacts. In the container space, you will see a lot of open source tools (among others) available that provide security for your end-to-end CI/CD pipeline. These tools can also benchmark your security against standards and best practices such as CIS security standards.

Configuration management

Cloud configuration checks are really important when you are continuously deploying to a hybrid multi-cloud environment. Businesses don't want to leave this down to a developer's wisdom or chance but ensure the environment is set up correctly instead. This includes checking that all the infrastructure, services, IAM (roles and access), and other service configurations are set correctly for the application. All clouds provide tools or scripts that you can leverage in your automation to perform these configuration checks. This confirms that you have secured your storage, compute, and network as well as your service on the cloud correctly. Another opportunity to shift security left is to check for certificate expiry before discovering it through an application outage. Data vulnerability and cloud configuration checks may also be automated as part of every release.

Security Orchestration, Automation, and Response

Security Orchestration, Automation, and Response (SOAR) is a core part of automation – identifying the risks, integrating the data that needs to be monitored, detecting the key event of interest, and being able to respond to it. Enterprises have many security tools in their landscape. Threat information from these various sources is one place for analysis. This calls for leveraging open standards such as the *Open Cyber Security Alliance* (https://opencybersecurityalliance.org/) and building a standards-based open ecosystem where cybersecurity products can share threat information without the need for customized integrations.

Integrated security and continuous compliance

Enterprises that have to meet industry and government regulations are taking a compliance-led approach to moving their workloads to the cloud. Clients have to deal with multiple certifications and compliance as separate efforts. When you have several controls to put in place and verify, it makes sense to automate the whole process. For the same reasons, the cloud providers offer a centralized facility such as the security and compliance center where clients can centrally manage compliance with organization and regulatory guidelines. In the cloud, it is possible to predefine groups of controls as profiles and use the results as a report for audit evidence and continuous compliance. To gain trust in the cloud, it should provide all the evidence and controls to meet industry-specific compliance and security requirements, specifically in industries such as banking, healthcare, and government.

Zero-trust architecture and security models

This is one of the most significant trends related to cloud security. This model considers a different model to addressing security compared to the traditional approach. More driven by multi-cloud scenarios, this is a new approach where the traditional perimeter-based protection with firewalls is changed to context-based access. This also supports one of the other macro trends that companies are trying to address – secure and remote access for the employees to enterprise resources.

The important aspects of this is you don't trust anything – people, processes, technology, networks, computes, or storage – until it proves that it is trustable. Some of the capabilities you need to build a security system based on zero trust are adaptive identity, context-specific and policy-enforced data security, policy-driven access control, and secured zones.

Several companies have built systems and platforms on top of the zero trust principle to establish integrity and trust levels explicitly. This is based on an organization's risk threshold and tolerance to provide access to assets and data systems. This requirement has been amplified during COVID times when most employees are accessing corporate resources from different places. Thin virtualized apps and workspaces are likely to back in demand to meet the security needs of employees logging in through the internet to access enterprise applications.

NIST has identified this requirement to access cloud resources from anywhere, on any device, and to expect a reliable and secure experience as able to be achieved through a zero-trust security model. The `NIST publication` discusses the various aspects of this model focused on protecting resources (assets, services, workflows, network accounts, and so on) and discusses general deployment models and use cases. Industries such as government and defense drive confidential computing and multi-level security requirements, which intersects with the overall zero-trust approach.

Summary

In this chapter, we learned about what the cloud is and the different deployment, delivery, and consumption models of the cloud. Enterprises recognize the business value of the cloud and how a hybrid multi-cloud approach is emerging as a key ingredient in successful digital transformation journeys. There are major observable changes across how applications are built, data is managed, integrated, and operated at scale. Security is a cross-cutting concern in all these areas.

All the security components and architecture required for this digital transformation journey needs to be carefully rethought – identities, networks, apps, data, integrations, devices, and analytics for cloud-based workloads. We need consistent, repeatable methods of architecting, designing, and integrating security for hybrid cloud applications to provide context-based access to resources based on a zero-trust model. However, this is a complex topic that needs to be discussed across various areas for a hybrid multi-cloud infrastructure.

We can look at the benefits of taking a pattern approach to break down this complexity and address the problem context. By specifying right architecture building blocks for the context, we can build reusable solutions as the basis for providing effective security for digital transformation journeys.

In the next chapter, we will learn how to work with patterns better, how to use them, when, why, and what trade-offs to consider. We will cover the security architecture, domains, roles, and responsibilities that we need to know.

2
Understanding Shared Responsibility Model for Cloud Security

As your business undergoes digital transformation with the foundation of multi-cloud, it is necessary to rethink your security architecture and methods of achieving continuous security. This approach to secure digital transformation is different from the past methods taken to secure traditional monolithic enterprise applications. In place of securing applications that were running inside their own data centers, enterprises need to deal with a heterogeneous ecosystem. For applications modernized and migrated to the cloud, there is a need to provide the assurance that the cloud is a secure, trusted platform. These applications also need to securely integrate with existing applications. Addressing hybrid cloud security concerns requires a strategic approach that's aligned with the enterprise's strategic objectives of a risk-based approach assuring total data protection and continuous security.

We will cover the following topics in this chapter:

- A strategic approach to cloud security
- A shared responsibility model for cloud security
- Cloud security domains
- A pattern-based and zero-trust approach to addressing hybrid cloud security

A strategic approach to cloud security

A strategic approach to cloud security should be driven by a governance and control framework created with the enterprise's risk appetite in mind.

Enterprise security teams led by the **Chief Information Security Officer** (**CISO**) define the policies and controls, as well as managing them against the IT budget. They leverage an IT **Governance, Risk, and Compliance** (**GRC**) framework that defines the policies and assesses the controls in place to meet the audit and compliance requirements. The audit and compliance set are based on industry needs and regulatory requirements. The risk management part of the framework continuously assesses the effectiveness of the controls against the business goals. The following diagram shows the GRC control framework, which includes the policies, controls, audit, compliance, and risk management:

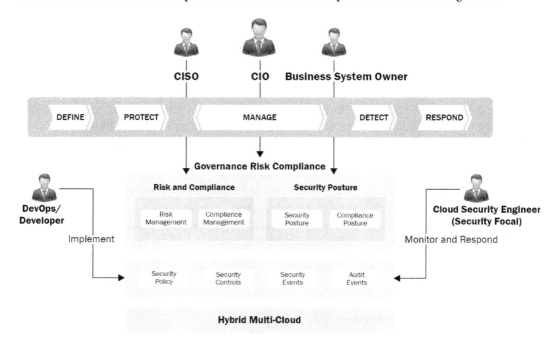

Figure 2.1 – GRC for the cloud

This also includes threat management and the continuous monitoring of multi-cloud environments. These controls need to be quantitatively and qualitatively measured to ensure continuous security. Achieving continuous security involves the cycle of defining, protecting, detecting, and responding on an ongoing basis. The leadership team manages this cycle against the business goals.

The cloud security strategy should also be guided by a set of principles that include but are not limited to the following:

- **Shift left security for a secure by design approach** – Ensure that security is included early on in the cycle, at the stage of conceiving the solution architecture and design. Currently, security is included toward the end at the time of deployment and operations. Addressing security issues needs to begin at the start of development, not at the end. A change to this approach is recommended.

- **Simplified and automated security management** – It is recommended to keep the policies and controls consistent across all environments. A centralized security and compliance center to define, enforce, and monitor security across hybrid multi-cloud environments is recommended. Build security solutions that are easy to consume and manage through automation. This will also help reduce the cost of security from an operational perspective as well.

- **The principle of least privilege** – This principle from the traditional software development model is relevant to the cloud world as well. Security privileges should be just enough to perform the activities. There is a need to continuously keep checking to ensure that there is no overprovisioning of privileges to any user or system.

- **A cloud-first approach** – It is recommended to leverage the controls that are available natively in the cloud to secure the workloads. Doing so means that security services can be consumed by the developers, similar to other cloud services. This consumption model also helps developers to easily secure their application without being security experts. For the security capabilities needed, the model is to look at the cloud first and then build on top of it with other partner products or technologies. The security focal for the cloud application can additionally leverage an existing homegrown or partner capability to perform end-to-end security management.

- **Complete data protection** – This principle recommends taking a data-centric approach to security. This involves three steps. The first one is to do data classification and understand the different types of data in the enterprise and define the protection needed for each type of data. The second step is to protect classified data based on security and compliance requirements. The protection would involve building defense-in-depth patterns supporting various access levels for different datasets. The protection implements security controls that ensure data is protected at all times. Finally, data access monitoring confirms that there is no unauthorized use and enables rapid corrective action if a breach or anomaly is seen.

- **Zero-trust models** – This is a set of emerging principles that can guide an enterprise security strategy. In this model, we switch from traditional perimeter-and content-based security to context-and workload-based security. This model moves from assumed trust to a zero-trust approach, where everything, such as users, devices, networks, applications, and data, is treated as untrusted elements. Trust must be established and verified instead. Verifying the access to provide to the resources is based on the context and the business needs. This model assumes a security breach can happen anytime and looks at ways to continuously improve the security posture.

A shared responsibility model

Depending on the cloud consumption and delivery models, the division of roles and responsibilities will vary. As discussed in the following diagram, the flexibility of each model varies, as does the responsibility for ensuring the security of the components. You will see that when we go from IaaS to SaaS, the responsibility of the cloud provider increases with consumer flexibility decreasing. It is necessary to understand the overall stack and what is being consumed from the cloud. Accordingly, we have to draw the line to determine what aspects would be taken care of by the provider and what needs to be covered by the consumer. Security of the overall cloud solution is a joint responsibility:

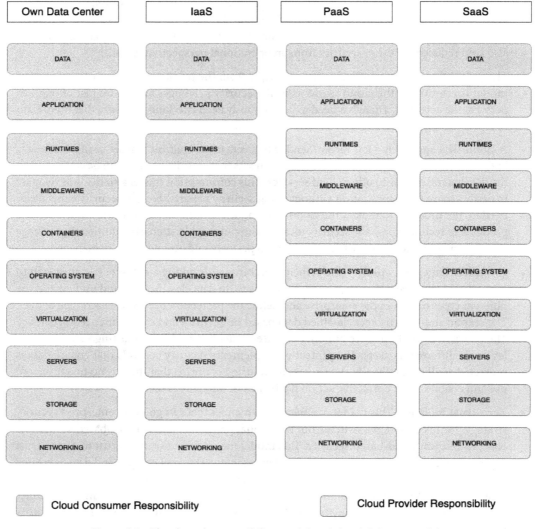

Figure 2.2 – The shared responsibility model and cloud delivery models

An IaaS model

In the case of IaaS, the cloud provider handles the traditional in-house responsibilities of the enterprise to provide storage, the compute, and network services. This includes the operational management of the physical facilities and the infrastructure (network and servers hosted in this environment). The enterprise consuming the IaaS will be responsible for the architecture, deployment, security, and operational management of the workload deployed on top of the infrastructure. Again, where you draw the line to divide the roles and responsibilities will depend on the service and contract with the cloud provider – for instance, the roles and responsibilities for bare-metal servers and those for virtual servers could be different. In the case of a bare-metal server, enterprises have exclusive rights to dedicated physical servers. They can install operating systems of their choice or bring their own, including bringing their own virtualization strategies and virtual appliances. This makes it easy to extend the enterprise's on-premises virtualization strategy to the cloud, independent of our virtualization options. So, with IaaS, similar or flexible options to what enterprises have in their own data centers are available. However, certain tasks will need a cloud provider's support, such as adding physical memory or swapping hard drives through a documented change management process agreed upon with the enterprise.

If an enterprise is consuming compute services such as virtual machines, the management of the underlying (physical) host and hypervisor including its configuration and patching lies with the cloud provider. The hypervisor is only accessible over the private management network maintained by the cloud provider. It is the cloud provider's responsibility to monitor the hypervisor's health and security including applying critical patches and upgrades to ensure the health of the virtual machines on the host. This might involve migration of the virtual services across hosts to meet the security and health requirements or support maintenance of the underlying physical server. The maintenance and update of the software and operating system running inside the virtual machine shall be the cloud service consumer's responsibility.

Similarly, in the case of a network, the various segments would include public, private, and management networks, and the private backbone for the data center. Cloud providers manage the configuration and management of the physical devices including the switches and routers that provide network structure and connectivity.

In the IaaS model, the provider therefore takes care of the security of the physical environment, and the data center, and makes the servers available for use by consumers. The security of the workload and managing access to it is the consumer's responsibility.

A PaaS model

When it comes to PaaS, the provider will provide the middleware services. These include capabilities such as messaging, a database, a cache, and functions made available as a service. In this case, the provider takes care of end-to-end security for the platform services. Designing the threat management for the platform, and taking care of patching the middleware, as well as the underlying server and system infrastructure, is the responsibility of the provider.

The PaaS consumer is responsible for the security of the application and data built or run on top of this platform. Security for any sensitive or **Personal Identifiable Information (PII)** data that the application is collecting, processing, or storing in the platform will be the responsibility of the cloud consumer. Meeting the regulatory requirements from the data and application perspective will also be the cloud consumer's organization's responsibility. The provider organization can help provide necessary supporting evidence or the provider can be contracted by the consumer to manage security or assist with regulatory compliance and reporting tasks.

A SaaS model

In the model, the provider takes on all the pain of securing the service from end to end. Right from infrastructure, middleware services and the application or workload security are all managed by the provider. The threat, vulnerability, and patch management of the entire stack is the responsibility of the provider. Clients get the true gain of a pay-as-you-go model with SaaS without having to worry about any of the security operational aspects. But the security of the data stored within the SaaS service would still be the responsibility of the cloud consumer. Continuous data protection and managing the keys used for encryption of the data will be the consumer's responsibility. Also, the operational aspects related to data access monitoring and protecting the IP are the consumer's responsibility.

The shared responsibility model

Enterprises need to define and establish the responsibility matrix related to how they are going to implement the strategy for meeting their cloud security requirements. Cloud security is a joint responsibility between the cloud provider and cloud consumer organizations. As we discussed in the previous chapter, there are several actors typically involved in building and operating a cloud solution, their roles, responsibilities, and the relationships between these actors. There are specific roles responsible for executing the strategy. Once an organization has started with the cloud, then some typical actors are involved in the design as well as day-to-day operational aspects of cloud security. These roles across the cloud providers and consumer organizations include the following:

- **Chief Information Officer (CIO)** – Defines the high-level information security and compliance policies for the enterprise. This role also communicates with C-level executives on the critical decisions concerning the technological strategy and component selection.

- **CISO** – Puts in place the framework to implement the policies for the enterprise. We also see a variation of this role as a **Business Unit Information Security Officer (BISO)** who is responsible for implementing the policies and controls for their business unit or the systems under their control – middleware, servers, architecture, and the technologies and tools to use for this. The CISO needs to understand the security management and compliance aspects, as well as the applicable standards leveraged as part of the implementation.

- **Security focal** – This role executes the plan created by the CISO and ensures the application or workload level security. The tasks include performing security scans, setting up tools, monitoring results, investigating risks or issues identified, opening tickets for fixes, and collaborating with development or operations teams. The tasks also include vulnerability management – every time a patch is released, they need to check whether it is relevant, install (or assign) it, and make sure it is done on time.

- **DevOps/SRE** – In teams where there is no dedicated security focal for the projects, the DevOps professional or the **Site Reliability Engineer** (**SRE**) works on the tasks to implement the recommendations and reports status back to the CISO or security analyst.

Security and compliance requirements don't change when you move your workloads to the cloud. A hybrid multi-cloud application involves many components across the IaaS, PaaS, and SaaS stack consumed from different clouds. Shared responsibility is a matrix that defines the roles and responsibilities of the provider and consumer. This captures the security obligations and the accountability of each stakeholder. Each participant and the roles of the cloud provider and cloud consumer organization are responsible and accountable for different aspects of security. The teams must work together to ensure nothing falls through the cracks or is left unattended. The shared responsibility model and management of risks and compliance could become the core deciding factors when planning multi-cloud adoption.

This responsibility assignment matrix is typically captured in what's called a **RACI model** (`https://en.wikipedia.org/wiki/Responsibility_assignment_matrix`). This matrix defines who would perform what task and the activities are grouped under a role:

- **R = Responsible** – This role performs the task or does the work to complete the task. A RACI table is incomplete without the responsible role filled in, while other roles may be optional.

- **A = Accountable** – This role ensures that the task is completed and is answerable to authorities on any queries on the task. The role arranges for the environment and prerequisites required for the task to be completed. Work allocation and the delegation of tasks are also the responsibility of this role.

- **C = Consulted** – Opinions are taken here, or this role brings in subject-matter experts for making decisions, as well as determining best practices.

- **I = Informed** – This role is generally kept in the loop or informed about the progress of the task. This communication would be more from a business, technology, regulatory, or compliance perspective.

The RACI matrix for a secure solution should cover all components that need to be secured and call out the responsibility of the users involved to bring together the roles and responsibilities into a single page. This could be done in two ways, either taking a resource-centric approach or taking a security activity-centric approach.

An example RACI matrix for a resource-centric approach is given in the following table for a PaaS model. We take the service as the cloud resource and list out the areas to be covered for the end-to-end security of the service. Then, the responsibilities are called out for each aspect to be covered to secure the resource:

Use Case / Area	Service Consumer	Service Provider
Infrastructure provisioning	I	R, A
Security for infrastructure		R, A
Cluster administration	I	R, A
Service administration	R, A	C
Security of application / data	R, A	C
Observability and security monitoring of app	R, A	C
Observability and security monitoring of infrastructure	R, A	C
Observability and security monitoring of dependent services	I	R, A

Table 2.1 – A resource-centric RACI model

An example RACI matrix for an activity-centric approach is given in the following table. In this approach, we take the service as the cloud resource and list out the areas to be covered for the end-to-end security of the service. Then, the responsibilities are called out for each aspect:

Security Control / Domain	Description of the activity	Provider	Customer
Physical / environment security	Security of physical and data center facilities		
Identity and Access Management	IAM roles and implementing separation of roles		
Event logging and audit events, protection/ retention of logs	Observability tools and controls depending on the cloud resource and contract		
Availability monitoring	Key parameters of service to be monitored		
Network isolation	Firewall/security zoning		
Edge protection	Web application firewall and perimeter protection		
Data processing and protection	Service data, PI data that is collected, and how it is used.		
Vulnerability management	Responsibilities across all layers in the stack infrastructure, dependent services, and client application		
Configuration management	Best practices provided by the provider and implemented by the consumer (based on the layer)		
Threat management	Identify the threats, risks, their detection, protection, and response		
Patch management	Proactive and reactive patching of components across the infrastructure, middleware, and application		
Incident management	How to respond to the security events		
Personal security/ training	Individual safety as well as education for personnel involved to maintain the security of the environment		
Business continuity/ disaster recovery	What elements will be brought back up by the provider and what components the consumer needs to take care of		
Compliance	Standards/certifications/regulatory compliance that the service needs to meet		

Table 2.2 – The activity-centric RACI model

Simply stating that the cloud provider will take care of everything when it comes to audit and compliance is not enough. The organization's GRC program has to clearly call out the information that the cloud provider needs to meet the audit and compliance requirements. These continuous compliance audits could be done by third-party auditors and the enterprise needs to collaborate with the cloud provider to provide evidence of the best practices for the end-to-end workload compliance requirements. This could be in the form of third-party-lead risk assessments, audits, and control reviews to ensure that the underlying foundation layers meet the security and compliance requirements and are managed to the highest standards. It's important that the various responsibilities are clearly documented in the cloud provider contract and are subject to audit. And the business and IT system owners need to have a good understanding of who is responsible for what security tasks. This also needs to include who does what in the case of a security incident.

The collection of these audit records or evidence may be automated for different cloud resources and services. The cloud provider can then share this material on demand with the consumer organizations through a web portal and API. This gives full visibility into the efforts and management of the environment by the provider. The enterprise consuming the cloud resources can import this information into their GRC dashboard or tools to make it easier for their own security and compliance needs.

The activities to be performed at each phase of the cloud project need to be planned jointly by the cloud provider and consumer. The phases would include project kick-off, engagement planning, design, development, testing, and operations.

Cloud security domains

Addressing hybrid multi-cloud security is a multi-layer, multidimensional problem to be solved. For an end-to-end security model, we need to understand each of these security domains and define the dimensions for each of the domains. We should be able to specify the security controls or processes for each domain and how to operationalize them to achieve continuous security. This model takes into account the need for a **continuous integration** (CI) and **continuous delivery** (CD) model in the cloud-native world. The security organization also needs to be aligned with this culture.

The following diagram shows the various security domains to be considered from a multi-cloud perspective, as well as the capabilities needed for security operations to support the CI/CD model for the cloud:

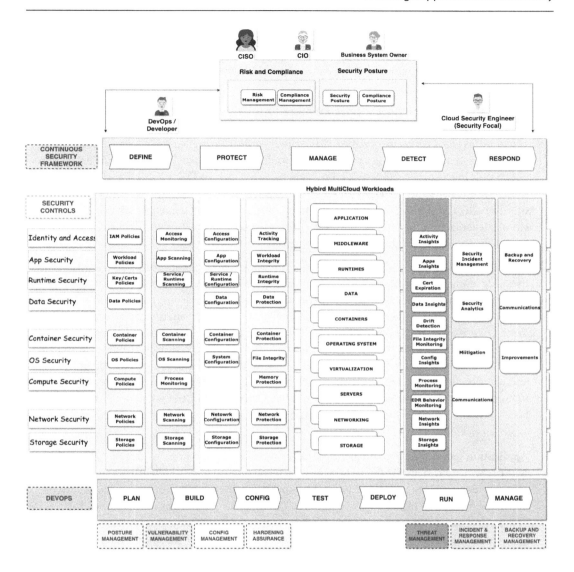

Figure 2.3 – Cloud security domains and components

The following are the domains to be considered:

- **Identity and Access Management** (**IAM**) – Identity is at the edge of all solutions. The IAM domain includes authentication, authorization, and audit. Authentication identifies the user or system and provides access to cloud resources, services, and applications. Authorization determines the level of access and operations that can be performed on the resources. IAM needs to be applied across all layers in the stack – the infrastructure, middleware application, and data. In multi-cloud environments, it needs to account for the different types of users and systems – including cloud users and application users. The aspects of managing authentication and access policies for cloud users is called **cloud IAM** and is called **application IAM** for an application.

- **Infrastructure security** – The infrastructure security domain includes providing a secure compute, storage, and network. Compute resources include bare-metal servers and virtual machines, as well as containers provided as a cloud service. The network is another critical aspect of the infrastructure. Securing the network would involve network isolation and protection and providing secure connectivity. Building secure zones and monitoring and responding to network threats such as **Distributed Denial-of-Service (DDoS)** are also requirements in this domain. There are different types of storage such as file storage, object storage, and block storage leveraged in the cloud ecosystem. The patterns for protecting the data stored in these storage types are part of the infrastructure security domain.

- **Data security** – This encompasses the aspects of how we protect data at rest, in transit, and in use. This leverages encryption standards, technologies, and tools, along with key management. In a hybrid cloud world, key management patterns are an important consideration, as they provide the consumer the ability to be in control of their data through patterns such as **Bring Your Own Key (BYOK)** and **Keep Your Own Key (KYOK)**. The required steps to protect highly sensitive data, such as encrypting it inside the application as well as while in use, are captured as data protection patterns. Interservice communication requires mutual authentication and securing communication channels and pipelines. This domain also requires certificate management capabilities that address the protection of data in transit, as well as providing better visibility into certificate life cycles and proactively managing certificate expirations to avoid service outages.

- **Application security** – This domain deals with ensuring the applications are free of vulnerabilities and protecting the application. Modern applications include applications that leverage modern technologies to provide a unique experience to end users. Enabling the secure use of these exponential technologies, such as IoT, blockchain, serverless functions, and the use of quantum computing, falls into this security domain. This includes applications that can be accessed through different channels such as the web, mobiles, or kiosks. Developers need to understand the risks of cloud resources being made available to end users through different application channels and devices. The steps to implementing authentication, the verification of channels and devices, and risk-based authentication are part of application security. Digital transformations built on the cloud also have an increased demand for a business platform that is enabled through the use of APIs. Building and scaling secure APIs is part of the application security domain. Security for the integration infrastructure, enabling connectivity with other applications, also needs to be included in this domain.

- **DevOps** – This domain includes the planning, building, testing, deployment, and management activities of software development and operations. The threat modeling aspects to identify malicious actors and the impact of their actions on the system start in this domain. This includes the proactive identification of threats and attacks that can lead to security incidents or loss of sensitive information. The team needs to plan for secure programming, configuration, and integration as part of the DevOps pipeline. Incorporating security testing of the code (static and dynamic) as part of the CI/CD process is also part of this domain.

- **Cloud Security Posture Management (CSPM)** – We need to discover the risks and continuously monitor the threats to the cloud resources. CSPM provides the means to this visibility. To get this visibility across all the cloud and non-cloud environments, we need to collect logs and events across all layers or domains – identity and access, infrastructure, application, and data. A unified dashboard or console for centralized security management is a component of CSPM. This dashboard also integrates findings from other security tools on misconfigurations that can lead to security exposures. Through this dashboard, security focals can view the security posture of all the deployments. **Security Orchestration, Automation, and Response (SOAR)** extends CSPM by alerting critical incidents and orchestrating automated responses. CSPM also provides a way to proactively carry out vulnerability, configuration, and threat management through the analysis of the security data collected from all the other security domains.

A pattern-based approach to address hybrid cloud security

Patterns are successful solutions to a repeating problem. As we understood in the previous section, the security domain is vast and includes multiple dimensions. It is difficult to find the skills and talent that can rightly address security challenges efficiently across all these domains. It is also difficult to find experts, security architects, and engineers who are knowledgeable across all these security domains. Hence, the recommendation is to take a pattern-based approach to solve the challenge of *how* to implement hybrid cloud security.

A security pattern provides a reusable solution to a generally repeating problem and provides guidance on how to design, develop, implement, and operate it efficiently.

A sample security pattern template is discussed in the paragraphs that follow. This template definition follows the details defined by the *Architectural Patterns* work published by The Open Group (`http://www.opengroup.org/public/arch/p4/patterns/patterns.htm`).

We will use this template to capture the problem context and the solutions across various security domains.

The following is the sample pattern template:

- **Name**: A short name that captures the essence of the security problem and solution.

- **Problem**: Describes the problem context with more details on the issues to be solved and the objectives to be met. This captures the functional objectives or business goals to be met by the solution.

- **Context**: This captures the context for the problem to be solved. Includes the pre-conditions, forces, constraints, and qualities to be met by the solution. The forces, constraints, and qualities describe the non-functional requirements to be met. This includes but is not limited to aspects such as the ease of construction, reliability, performance, scalability, throughput, availability, correctness, effectiveness, resiliency, and serviceability.

- **Solution**: This describes the components that make up the solution and how they interact to deliver the intended goals. The solution or blueprint captures the static as well as the dynamic behavior of the solution including the systems, actors, and their collaborations. With a schematic diagram, the solution section details the various components and their interdependencies. The solution section also discusses the alternatives with relevant trade-offs and the rationale for deciding on a particular option versus another within a specific context. At a high level, it captures the post-conditions after the pattern is applied.

- **Known uses**: This section will share an example of how the pattern is applied and the resulting context. This will share how the pattern can be implemented with various cloud providers such as AWS, Google Cloud, Azure, and IBM Cloud.

- **Related patterns**: Patterns do not exist in isolation. Multiple patterns may need to be combined to resolve a complex problem. This section will also capture alternative patterns and a dependent predecessor, as well as successor patterns however applicable.

- **Known uses**: Known applications of the pattern within existing systems, verifying that the pattern does indeed describe a proven solution to a recurring problem. Known uses can also serve as examples.

The Open Group template has the forces as a separate section of the pattern template. To keep it simple, the forces, qualities, and constraints applicable to a scenario are captured as part of the context section in this book. This book includes the patterns across the various security domains and groups them under the following categories, listed as follows:

- IAM patterns

- Infrastructure security patterns

- Data security patterns

- Application security patterns

- DevSecOps patterns

- Cloud security posture management and response patterns
- Zero-trust patterns

Summary

We have discussed the *what* and *why* of hybrid cloud security in these first two chapters. Now, we need to know the difficult part of *how* to achieve hybrid cloud security. That is what the remaining chapters of this book will focus on.

In the next chapter, we will learn about the IAM domain further and the patterns to implement authentication, access control, and auditing for the cloud.

Part 2: Identity and Access Management Patterns

Cloud **Identity and Access Management** (**IAM**) is the core component for managing identity and access for the platform and cloud applications. Application IAM manages the identities and access for applications. This part will provide you with the reference architecture, best practices, and patterns to leverage for addressing identity and access.

This part comprises the following chapters:

- *Chapter 3, Implementing Identity and Access Management for Cloud Users*
- *Chapter 4, Implementing Identity and Access Management for Applications*

3
Cloud Identity and Access Management

Enterprises need a seamless way to provide access for their users to applications and resources on multiple clouds. **Identity and Access Management (IAM)** is the core component for managing identity across a hybrid cloud. This has two important considerations to manage and solve:

- How to manage identity and access for users to cloud resources, services, and platforms, which is typically referred to as Cloud IAM.

- How to manage identity and access for users for cloud applications, often referred to as IAM for cloud applications

IAM deals with identifying a user and cloud resources and managing access to these resources for the user, based on policies. IAM provides fine-grained access control to cloud resources and services under specific conditions. IAM policies define and manage permissions for each user to specific resources. IAM goals are more business-aligned than technical components. Enterprises need a robust and mature IAM capability to keep their costs at a minimum and that can be swiftly changed based on business requirements. In this chapter, we will take a deep dive into the patterns and best practices for addressing identity and access for users, leveraging security services, tools, and technologies provided by the cloud.

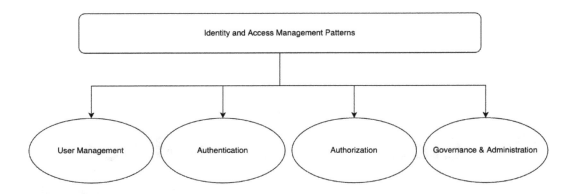

Figure 3.1 – IAM patterns

As shown in *Figure 3.1*, from an operational perspective, IAM patterns are covered under the following areas in this chapter:

- User management
- Authentication
- Authorization
- Governance and administration

IAM applications extend these requirements to cloud applications and their users. This will be covered in the next chapter.

User management patterns

Within cloud environments, you need to plan for providing and managing access to different types of users and identities. Users of the cloud include the following:

- **Admin users:** This set is the group of administrative users that typically includes product owners, managers, and team leads of cloud applications. They also manage the other users involved in an application, such as the developers and operators. These users typically require complete access to the application and platform development environments. They also need visibility into each cloud user's activity on the cloud. These user accounts are authorized to view sensitive information and execute actions that can have a wider impact. So, an increased level of protection, auditing, and governance is required for admin users.

- **DevOps users**: This set of users includes developers and operators, cloud application developers, and platform developers who create, maintain, and delete applications. Additionally, this type of user has the need to create and consume additional services from the cloud. These activities of this user set are also subject to audit, as they are typically authorized to read sensitive information and can update applications and data. So, the access of developers is restricted to only specific project areas or environments in the cloud, based on the requirements.

- **Application users**: These are users of cloud-hosted applications. These applications can be web or backend applications. Users could also include other applications or systems that consume APIs from the cloud. Depending on the user and the requirement, the identification and authentication model is selected. IAM for this set of users will be discussed in detail in the next chapter.

The user management patterns cover the problems, contexts, and solutions for the following key goals:

- Managing a user:

  - Single-user onboarding/self-registration

  - User provisioning

  - User de-provisioning

  - Options to manage user attributes and credentials

- Managing groups:

  - Bulk user onboarding

  - Group creation

  - Group deletion

It is also critical to protect cloud users from identity theft and privacy abuses, limiting the potential for and impact of a cyberattack, and streamlining numerous operational processes. Identity management is the process of managing information used to identify users. From a cloud user's perspective, they would want to use a single identity and access any cloud resources for the development, deployment, or management of their applications on the cloud.

Registration pattern

We will look into the following example.

Problem

How do we onboard a new user to a cloud platform?

Context

This involves how to create or register users, individually or in bulk, to use the cloud and related resources. The stages under identity management include requests for onboarding, provisioning the identity, securing the approvals if needed, and propagating the identity to the target systems as needed. Communicating the various stages to the relevant parties through notifications is another key aspect to be addressed under identity management.

Solution

Before any user can access a cloud service or resources, they have to complete an onboarding or registration process. The following figure shows the pattern a user can use to register for cloud usage through a self-registration page:

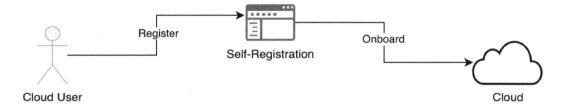

Figure 3.2 – User self-registration

An individual cloud user, such as a developer working on a personal project or a small company, can self-register as a new user for the cloud through the registration page provided by the cloud provider. Then, they can use their ID to access cloud services and resources. The registration process creates a user and associates them with an account. The user acknowledges the service agreements listed by the cloud provider. The cloud provider acknowledges the personal data collected and creates a new user. The credentials the user has shared can be used for subsequent logins to the cloud to use the service or access the resources, as shown in the following diagram:

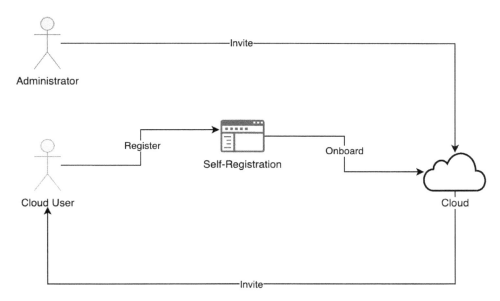

Figure 3.3 – Admin-initiated invite and self-registration

Each cloud user will be associated with an account or organization as part of the onboarding process for billing, contracting, and support purposes. The owner of the account also needs to be created for user administration purposes. The account owner will have privileges to create other users for their account. As shown in the preceding diagram, an organization or account owner can create new users by specifying a temporary password. The user then receives the notification and has to complete the process, entering a permanent password and other attributes required to complete the registration.

Identity federation pattern

We will look into the following example.

Problem

How can an enterprise onboard several of its employees to the cloud?

Context

The self-registration pattern is not a feasible and scalable option to onboard a large number of employees to the cloud. Identity management is crucial to an organization, and it needs to be highly available and scalable, since identity is at the edge or the entry point to every application or microservice in the cloud. It can influence the productivity of your employees and the security of your organization.

The other key challenge is that the user life cycle management needs to be very dynamic. The user can change departments and be given new roles and responsibilities. If the user is transferred within the organization or leaves it, the access and permissions need to be reviewed and updated. The review might involve an approval workflow process, and the ID is suspended or deactivated on time.

Solution

Enterprises typically use an on-premises directory such as an active directory, **Lightweight Directory Access Protocol (LDAP)** directory, or cloud directory as a centralized identity store. Onboarding enterprise users in bulk to the cloud is achieved by connecting their internal or external user directory (an **Identity Provider (IdP)** directory) with the cloud. This pattern is called **identity federation** and is described in the following diagram:

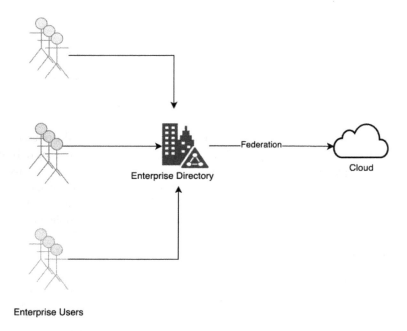

Figure 3.4 – Identity federation

This pattern is supported by the underlying **Security Assertion Markup Language (SAML)** or **OpenID Connect (OIDC)** technical protocol. Federation enables **single sign-on (SSO)** to cloud accounts and applications. In this pattern, clients can use their own enterprise identity with an associated password to log in to the cloud. For enterprises, this makes it convenient to centrally manage federated access to multiple cloud accounts and business applications. In this model, they can also manage and assign access to cloud resources based on the group membership in the IdP directory.

Cloud identity pattern

We will look into the following example.

Problem

How do we manage the provisioning of users centrally across cloud and non-cloud environments?

Context

Certain organizations may not have a scalable user registry to support a federation pattern. In some cases, the directory might be fragmented across different departments or might not support underlying technical protocols for the identity federation pattern. Also, the existing user directory may not support user profiles centrally. The current implementation might store user profiles separately along with applications.

Solution

A cloud identity solution is used to implement automated user provisioning for cloud-based applications. This solution, also called **Identity as a Service** (**IDaaS**), enables enterprises to spend less on enterprise security by relying on a centralized directory to deal with the provisioning and management of users across cloud and on-premises applications. A cloud identity allows users to work from any location and any device, providing SSO with one set of credentials to multiple applications.

The benefits of a cloud identity solution or IDaaS include the following:

- A centralized identity management system that hosts user profiles and credentials, making it easy to manage access to applications at one place and having visibility through a single dashboard.

- Cloud identity is usually backed by cloud directory services that support the storage of user profiles and associated credentials in the cloud. The user details can be accessed through a LDAP and shared across applications or organizations.

- Cloud directory services securely manage user profiles and their associated credentials and password policy inside a cloud environment. A directory service within a cloud means that applications hosted on the cloud do not need to use their own user repository or existing (legacy) user repository.

- A cloud directory service facilitates the sharing of network capabilities with a complete spectrum of users for secure collaboration, such as a central identity hub for connecting with suppliers, partners, and customers. This pattern can be leveraged to create bulk users for the cloud. Cloud directory management APIs can be leveraged to bulk import users.

- Bringing in cloud scale availability and scale for identity management to cater to spikes or on-demand capacity to address new cloud projects.

- Most enterprise-grade IDaaS implementations come with 24/7 support and centralized monitoring of the infrastructure, with overall increased productivity.

- Quicker onboarding of new users and projects, especially when an enterprise acquires another company or when additional business units need to be onboarded to the cloud.

- User records can be provisioned by or to external identity repositories by using identity management feeds.

- Cloud identity solutions can interface with many types of identity repositories, such as Active Directory, LDAP v3, relational databases, SOAP services, a message queue, and SAP. As shown in the following diagram, users can be automatically added to, modified in, and deleted from cloud identity services through integration with these other identity repositories by defining an inbound connection. Users can also be added to, modified in, and deleted from external repositories by using an outbound connection, as shown in the following diagram:

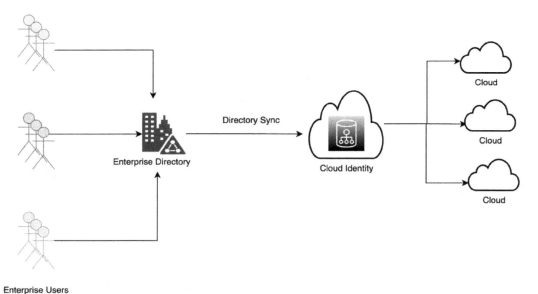

Figure 3.5 – A cloud identity solution

- Configuration can be done to enable identity data or fed to include all or specific attributes, groups, roles, and account information, to flow between the identity repositories and a cloud identity solution.

- Automation is set up in the form of workflows or assembly lines that include actions configured to execute specific identity feeds.

- A cloud identity solution provides a full user and application life cycle by creating, updating, removing, or suspending user profiles.

- Accommodates full app life cycle management by enabling companies to add or remove applications from their organization in a central location.

- Cloud identity comes with provisioning connectors that create users on target cloud and third-party applications. We can synchronize all or a subset of users to one or more supported apps.

- There are multiple patterns for an on-premises or internal user directory with a cloud directory used by the cloud identity solution. Synchronization makes a copy of the ID and forwards it to cloud identity management. On a scheduled basis, changes in the directory are pushed from on-premises to the cloud. Note that synchronization alone does not support SSO but only replicates the identities from one repository to another.

- With synchronization alone, a user will have to log in again to access cloud services or applications. A cloud identity, with automated provisioning or federation of the identity to the cloud, can support SSO. With a workflow to push the identity to specific applications, the user needs to log in to the cloud identity and then can use SSO to sign in to cloud platforms.

User group management patterns

We will look into the following example.

Problem

How do we manage IAM for users as a group?

Context

Enterprises need an easy way to create a user group. They also need to manage the membership of that group, such as adding and removing users. The administrator of the group should be able to change and update the attributes of the group.

Solution

In this pattern, as shown in the following diagram, admins can add or remove users from a group. Group attributes, such as names and other details, can be modified:

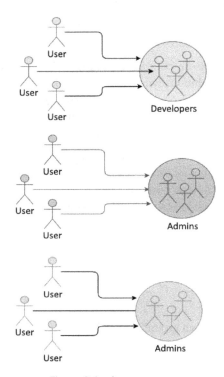

Figure 3.6 – A user group

A group makes access management easier. Once a group is assigned to a role, all the users in the group get the same access privileges defined in the role.

Service accounts

We will look into the following example.

Problem

How do we create and maintain an identity for services and applications?

Context

Like users, applications or services require unique identification. Based on an identity, a cloud admin can allow or deny access to specific resources and services for an application.

Solution

This is achieved, as shown in the following figure, through the concept of service identity or a service account:

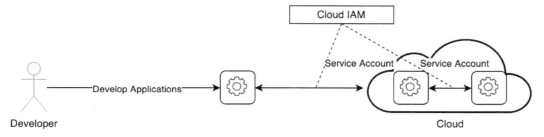

Figure 3.7 – A service account

A service account can be attached to any application or resource. Services IDs can also be added to a group, and one or more roles can be specified for that identity. An application or service uses this identity to interact with other applications, APIs, or services, or to invoke any API. A service identity replaces a user identity for application-to-service or service-to-service interactions. Typically, administrators and DevOps team members create the service ID as well as associated credentials, which are also referred to as the API key that is used for authentication.

User de-provisioning

We will look into the following example.

Problem

We need a robust and secure mechanism to remove a user's access to cloud applications and resources.

Context

When a user leaves a company or switches their role, they need to be de-provisioned or their permissions to specific projects or resources should be immediately revoked. The administrator also needs to ensure that the user's other identity-related details are removed, such as their password or login profile, access keys, **Secure Shell (SSH)** keys, and signing certificates, their **multi-factor authentication (MFA)** devices are deactivated, they are removed from groups, and their user profile is completely deleted.

Solution

Manual deletion or deactivation of a user can be done through an IAM console or by leveraging IAM APIs. As shown in the following diagram, user de-provisioning or removing cloud access for a user can be automated through integration with a cloud identity solution:

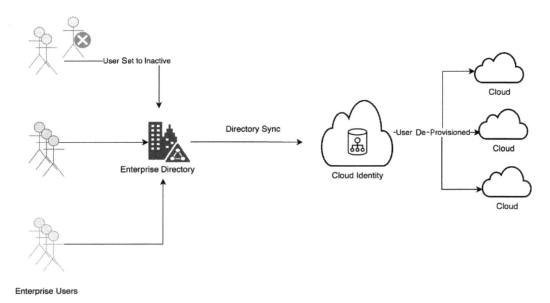

Figure 3.8 – User de-provisioning

The de-provisioning enterprise user workflow can be triggered from an HR system. This will ensure that cloud access to an employee who has left an organization is immediately removed. The process typically involves a workflow that takes approvals from the supervisors and different system owners. If the process isn't automated, there is a greater chance of a step in the workflow being missed. The periodic auditing of this process is also an important aspect to remove any unused or dangling credentials.

Authentication patterns

Authentication is an IAM function that establishes an identity when attempting to access a cloud resource. Within hybrid environments, you need to plan for authenticating different types of users and identities. You need a way to uniquely identify and authenticate a user to allow them access to a cloud platform. An authentication solution should also be able to authenticate users based on a range of identity providers.

Authentication functions recognize a subset or combination of the following identity providers:

- A cloud directory

- A social identity provider (such as Google or Facebook)

- An enterprise-hosted identity provider

- A cloud-hosted identity provider

Once logged in or authenticated by an identity provider, a cloud user should be able to use the identity context (IAM token) to access cloud runtimes or services without having to log in again. This is also referred to as an SSO requirement.

Similarly, the cloud user may have multiple active sessions with services on the cloud. The solution should allow the user to log out from each session separately or log out from all sessions using a single page or by running a single command (single sign-off). Single sign-off is harder to achieve because of the various client agents through which a user could be accessing resources or applications.

The authentication solution needs to support multi-protocol to enable virtually any IT resource to connect in their "native" authentication language.

Logging in with user ID and credentials

We will look into the following example.

Problem

How can a user log in to the cloud and access any resource, runtime, or service using their user ID and password?

Context

With authentication, a cloud user can log in to the cloud with their user ID and password.

Solution

As shown in the following diagram, cloud IAM allows users to sign in to the cloud using their user ID and password. IAM users can use the console to create, change, or delete a password:

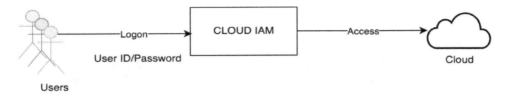

Figure 3.9 – Logging in with user ID/credentials

With the password, a user can sign in through the URL for their account.

The IAM administrator for an account can set the policies related to login, such as controlling how many times a user can try to log in before the account is locked. Also, as part of password policy management, the admin can enforce the need to have strong passwords that meet an organization's compliance needs.

Application access key or API key

We will look into the following example.

Problem

How can a user log in to the cloud through a **Command-Line Interface (CLI)** or by leveraging APIs using their ID but also create different credentials for each application or tool?

Context

A user needs an application access key for use with a CLI or their application when their user account is created, or soon afterward. For better security, the access key should be issued for only a specific number of uses or for a defined time period and duration. Users also need the capability to rotate their application access key at regular intervals or any time they feel it has been compromised.

Solution

A cloud user can also create an access key that they use to log in to cloud-hosted services or applications. This is mainly a requirement when using a CLI or logging in to the cloud through an application with a user ID. In this pattern, the cloud supports requests from the user for an access key. As shown in the following diagram, the user can request an application access key or API key:

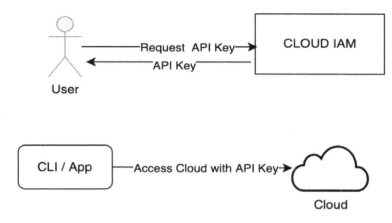

Figure 3.10 – Logging in with an API key

If a user wants to log in to the cloud through an app or command-line tools, then they can use an API key in place of a user ID and password.

SSH keys

We will look into the following example.

Problem

Users need a passwordless method to authenticate and log in to a cloud server from their client machines.

Context

Users need a safer authentication to the server, which is performed without passing a password over a network. If user IDs and passwords are shared over the internet, an enterprise runs the risk of interception by man-in-the-middle attacks.

Solution

Cloud users can generate SSH keys to authenticate and log in from their client machines to cloud servers or services. SSH keys help to identify an SSH server through public-key cryptography. The following diagram illustrates how a user uses an SSH key to log in to cloud IAM:

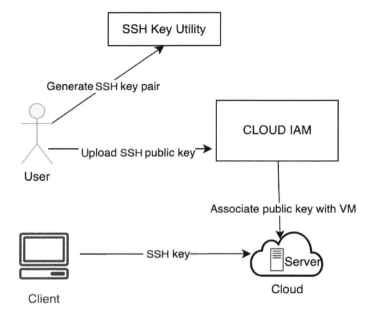

Figure 3.11 – Logging in with an SSH key

As shown in the preceding diagram, the steps involved are as follows:

1. Create an SSH key pair.
2. Upload a public key to the cloud.
3. Associate a public key with VMs.
4. Log in from a client using a private key.

The main advantage of SSH keys is that authentication to the server is performed without passing a password over a network. SSH keys prevent interception, man-in-the-middle attacks, or attempts to crack passwords through brute-force attacks.

Authentication with SSH keys adds additional IAM admin responsibility to keep track of active keys and their usage. If the keys are not deleted once a user leaves an enterprise or project, they can continue to access VM instances, which is a high-security risk.

SSO

We will look into the following example.

Problem

Once logged in, a cloud user needs to access other runtimes, services, or resources without having to log in again.

Context

Ideally, a user, once logged in to the cloud, should be able to access all services and resources that they have access to without having to re-enter credentials. With SSO, the cloud user can use one set of credentials to log in to multiple services and applications. This provides an improved user experience and helps in credential life cycle management, such as resetting a password and not having to remember passwords for multiple services.

Solution

SSO incorporates a federated-identity approach by using a single login and password to create an authentication token that can be accepted by multiple target applications. Once a user is authenticated through a centralized authentication server, SSO generates a browser-based encrypted session cookie. When the user logs into the cloud, cloud IAM checks for a valid session cookie. The user details are read from the valid session cookie, and the user is authenticated to the targeted cloud. The SSO pattern allows enterprises to enforce security policies in one place.

The following diagram shows how enterprise SSO facilitates the storage and transmission of user credentials to provide sign-in to the cloud:

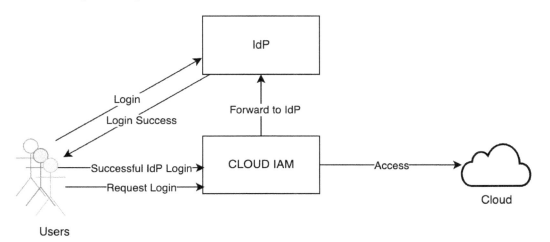

Figure 3.12 – SSO

The SSO pattern uses a standards-based authentication model. These standards include SAML and OIDC. The same standards apply to SSO for cloud applications as well. These will be discussed in detail in the next chapter. SSO combined with multi-factor authentication lowers the risk of security breaches.

Multi-factor authentication

We will look into the following example.

Problem

How do we provide an additional layer of security for authentication in addition to credential-based authentication?

Context

Enterprises require this extra layer of security to protect logins to cloud-based applications and lower the risk of security breaches – that is, enterprises need to implement authentication with more than a username and password. The username and password-based authentication method is not a strong method, as passwords can be stolen or cracked using brute-force attacks.

MFA makes the method stronger, with steps that make it hard for hackers or criminals to get access to cloud services or resources.

Compared to credential-based authentication, MFA provides a better way to secure digital assets and transactions over the internet through an additional layer of security. MFA also helps organizations maintain compliance with authentication processes and procedures. The MFA authentication model is shown in the following diagram:

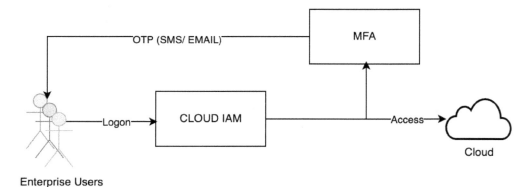

Figure 3.13 – MFA

MFA combines two of the following factors for authentication:

- **Something you know**: This could be a username and password

- **Something you have**: A mobile, USB, or a hardware security key or smart card to verify your identity

- **Something you are**: Biometrics data such as fingerprint, iris scan, or some other data that proves you are who you say you are

Adding this secondary factor to your username/password protects your privacy, and it's remarkably easy for most people to set up.

Typically, a **One-Time Password** (**OTP**), CAPTCHAs, or patterns are used as the secondary authentication mechanism along with credentials. Many enterprises use **time-based one-time-passwords** (**TOTPs**). One widely used implementation of TOTP is the use of a virtual authenticator app such as the **Google Authenticator** app with smartphones.

The usage of security questions is an alternative method for providing MFA. In this pattern, a user fills out security questions from a predefined list and defines the answer. At the time of authentication, the predefined security questions appear on the login screen, and by providing the defined answer along with the credentials, the user is authenticated.

In this advanced model, clouds support MFA with a chip and **Personal Identification Number** (**PIN**), which is a conventional method for the authentication of financial transactions. This authentication mechanism can be used for machines/services in a network to log on to the cloud. The asymmetric encryption technology uses public and private keys to encrypt and decrypt data with a chip and PIN mechanism.

Single logout

We will look into the following example – a **Single Logout** (**SLO**) is a solution pattern that is difficult to achieve.

Problem

A cloud user signed into multiple active sessions or services in the cloud needs to have a way to log out from each session separately or through a single command.

Context

The user could be logged into multiple cloud services and applications. In the absence of SSO, the user needs to log out of these sessions separately.

Solution

The SLO pattern is supported by certain clouds. While the web-based connected cloud services can be logged out of through this pattern, the pre-requisite is that the cloud supports SSO integrated with an IdP.

As shown in the following diagram, a typical logout request follows your typical SAML message structure, with an ID, lifetime data, and information about its origin and destination:

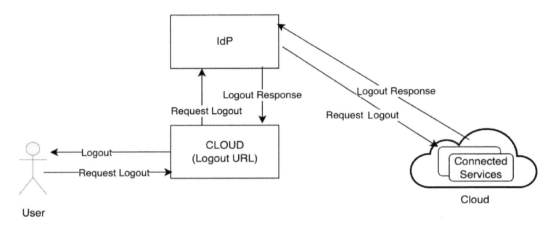

Figure 3.14 – SLO

This pattern follows the following steps:

1. A user initiates the logout process by clicking on the cloud URL.

2. The cloud terminates the user's session and triggers it by sending a logout request to the IdP.

3. Upon receiving a logout request, the IdP first identifies all connected cloud services that are part of the current session and performs the logout one by one.

4. After each connected service terminates the user's session, the IdP terminates the user session and sends the logout response.

5. The logout response typically includes a status code that informs the end user whether SLO has completed entirely or partially.

Physical authentication pattern

We will look into the following example.

Problem

How do we enable only approved employees with a valid business justification with physical access to a data center?

Context

Physical authentication is used by cloud providers to authenticate users needing access to a data center. The physical security of data centers is important. The implementation and usage of physical security are subject to governance, compliance, and audit.

Solution

Only approved employees with valid business justification are granted access. The principle of least privilege is followed while processing such access requests. The access is provided only for a timed window to restricted areas and the individual's activities are monitored. As shown in the following diagram, access cards (physical identity cards) and biometrics are used for authentication:

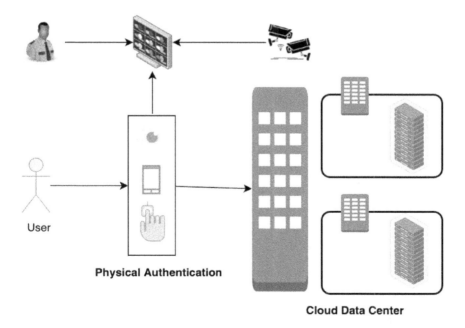

Figure 3.15 – Physical authentication

Biometric authentication is based on either fingerprint recognition, iris recognition, or face recognition, with one or more of those chosen. The physical security, implementation, usage, and governance policies of data centers are subject to compliance and audit.

Authorization patterns

The solution patterns for authorization are discussed in the following sections.

Access control pattern

We will look into the following example.

Problem

How do we control user access by determining their privileges and provide access to a heterogenous cloud environment consisting of services and resources?

Context

An administrator of a cloud for an enterprise should also be able to define common control rules and customize them as needed. There is also a need to enable granular access to specific cloud resources and services.

The problem to be solved is to authorize a user and provide specific access to cloud resources, services, and applications. The solution should manage access to cloud resources by different types of users as well as services.

The IAM administrator should be able to define and enforce security policies. The policy should apply the principle of least privilege, which gives users only the required permission to execute an intended action.

The solution needs to be able to grant fine-grained access control or a specific permission on specific services or resources for cloud users.

An effective authorization design should adhere to policy administration and offer an easy entitlement checking mechanism, for the enforcement point (such as a cloud service) to decide whether to provide access or not to a service or a resource that it manages.

Solution

Clouds achieve an authorization function by defining an access control model or pattern that consists of users, policy, and resources.

User and user group

As we already discussed in the user management patterns, there are different types of users for the cloud. Authorization needs to define the access control model for each type of user. Users can be either individuals or a group. Individual users can again be either a human user or a service account.

Policy

To define access for users and service accounts, you must create a policy that connects an identity to a resource by specifying one or more permissions for that identity. A set of such permissions is defined as a role.

So, a policy connects the user and resource through a role. A policy can specify a set of policy conditions under which access is allowed or denied for a user to a resource. The administrator who defines the policy can attach one or multiple policy conditions when the user is assigned a specific role for a cloud resource. A policy condition provides a way for access control decisions to be made based on a dynamically changing context, not just fixed rules. If a user is not assigned any role, the user does not have any privilege to access anything on the cloud.

Resource and resource groups

A resource is any cloud resource whose access needs to be managed by cloud IAM, which can be either infrastructure resources such as compute, storage, or networks, platform resources such as messaging, caching, or a data service, or application resources. A resource group is a group of related resources for a solution or a set that needs to be managed together. A cloud resource is uniquely identified by a cloud resource name. Clouds organize the resources in a hierarchy or inside a cloud account. A policy can be applied to a group of resources or at different hierarchical levels, and it is possible for resources to inherit policies.

The cloud access control model or pattern consists of three important elements, as discussed previously and shown in the following diagram:

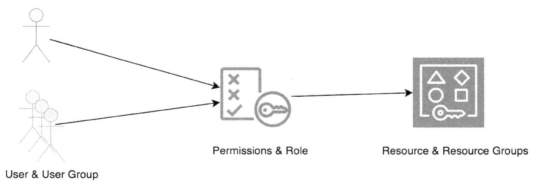

Permissions & Role   Resource & Resource Groups

User & User Group

Figure 3.16 – Access control pattern

A typical resource hierarchy defined by the clouds is given in the following diagram:

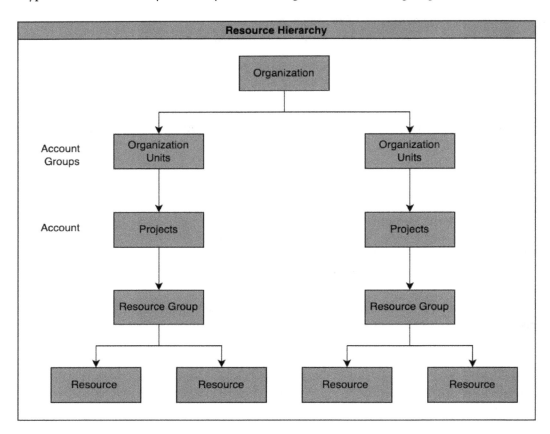

Figure 3.17 – The resource hierarchy

The resource hierarchy has multiple levels, starting with an enterprise or organization at the top. The next level is typically departments or organizational or management units. Some clouds also have the notion of account groups or folders at this level. Then, we have the account or projects level. From a cloud consumer perspective, each of these levels helps aggregate spending on cloud services and resources. This is mainly leveraged for billing and chargeback as well as rolling up the cloud expenses to the organization level. From an IAM perspective, the platform roles apply at the organization and account levels and are grouped under different areas. Services and related resources are created under an account or project. IAM access control provides cloud consumers with a single view of all the resources that a customer owns organized in a hierarchy, with access control applied to objects at each level.

Platform roles

A role is defined to specify exactly what actions (for example, create, read, update, or delete) can be performed on an object. Permissions defined at a cloud level or cloud platform level are called platform roles. Most clouds define predefined platform roles of owner, administrator, editor, operator, and viewer. An IAM administrator can select these predefined roles and assign them to any specific user or group. Platform roles or permissions are typically grouped into areas such as administration, IAM, billing, support, security, hardware, software, and network.

Service roles

Cloud services can define service-specific roles, such as reader, writer, and manager roles. For each object type, several default roles are defined to meet typical customer needs. Consistent role names are used across all cloud services, but each service still defines the exact actions for a role. Clouds also support the creation of "custom roles" to fit an enterprise's specific needs by combining specific actions to be supported for a specific service or resource.

The cloud can leverage any of the following access control patterns to enforce a policy for a user or resource:

- **Mandatory access control (MAC)**
- **Discretionary access control (DAC)**
- **Entitlement/task-based access control**
- **Role-based access control (RBAC)**
- **Attribute-based access control (ABAC)**

A cloud service typically invokes a cloud IAM API to retrieve the entitlement (allowed actions) for a user. The entitlement information allows a service to make an access control decision on a request for access to a cloud service or resource.

Governance and administration patterns

Governance and administration patterns cover the operational aspects of identity management – specifically provisioning digital identities for the cloud and applications, securing them, and ensuring the processes are rightly governed to meet audit and compliance risks.

Identity governance and administration pattern

We will look into the following example.

Problem

Enterprises need a way to report and audit all the user activities in the cloud. This will include reporting across user management, authentication, and authorization functions. Identity governance is a key aspect of cloud security. Without this capability, a badly provisioned user or user entitlement can put entire enterprise cloud resources at risk.

Context

Enterprises need to have IAM reporting and governance because of the following reasons:

- Reporting helps gain higher visibility of users' activities
- Analyzing IAM data helps to quickly identify possible security issues
- IAM reports help them to comply more effectively with audit and regulatory requirements

The key set of activities to be covered includes the following:

- User management reports:

  - User identities and group or role membership
  - A list of users with administration rights
  - Custom roles and users assigned more than default or special privileges
  - A user's login and password policies
  - API keys, service accounts, and their usage
  - A list of restricted users and restrictions, such as cloud portal access from specific IP addresses – for example, a customer's enterprise network

- A report on user credentials and authentication:

  - Listing all the credentials belonging to a user and their creation and update history.
  - A user's authentication history, including when they were authenticated, how (using what credential), and the result of the authentication. The report should contain all the users, all the credentials of each user, and the authentication history of each user.
  - User accounts and a credentials audit report.

- A report on authorization and roles:

  - List all the users and objects (account, projects, services, and resources)

- Admin can review when what access is provisioned for an user, all the actions that they can perform, and the history of when the access was granted through a role assignment

- List the events related to entitlement retrieval and the policy decision events for specific or all users/cloud resources

- Ensuring **Separation of Duties (SoD)** control, which prevents a user from having both the creation and approval rights for a task or ensuring that at least two individuals are responsible for a task

- A review of the assignment of roles to users and service accounts across projects and resource groups – for example, a developer having access to multiple projects or both development and production environments

Solution

All clouds have an IAM console that provides an integrated portal through which consumers can control and manage access to cloud services and resources for their users. While IAM admins can manage a user's permission on a cloud platform (platform roles) as well as role assignments for the services through a console, all the IAM management functions are available via an API. IAM provides a single dashboard through which consumers can get reports on users, authentication, and authorization. This pattern is discussed in the following figure:

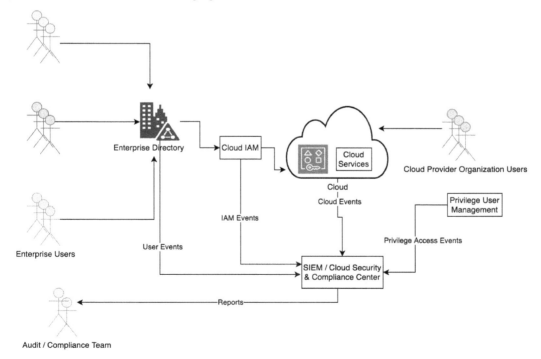

Figure 3.18 – Resource hierarchy

The *Who has access to what under what conditions?* report is often required for audit and compliance. This can be requested through the IAM console or generated by leveraging the cloud IAM APIs. In the case of enterprises that use a cloud identity solution or have federated their internal identities to use the cloud, data is compiled from reports from the cloud as well as the cloud identity solution. Identity governance activities go beyond reporting to take automated administrative actions. These actions can include identifying, checking for compliance and violations, as well as rectifying any issues related to or arising from the following:

- Identity life cycle management

- Password or credentials management

- Enforcement of the least privilege principle

- Role modeling, harvesting, and management

- Policy enforcement and SoD

- Certification for continued access

- Review of IAM configuration and settings

- Analysis of activity logs and user behavior

Audit and compliance are areas of focus. Both the cloud provider and the cloud consumer are required to work together to prove to an auditor that the controls put in place meet an organization's security policy. A risk officer can validate implemented controls for gaps from a regulatory and industry compliance perspective and report deviations. Compliance standards such as HIPAA, PCI/DSS, and NERC are mandatory for specific industries, and an automated report helps organizations efficiently demonstrate compliance to these standards.

Privilege Access Management (**PAM**) is used to lower a security risk and make compliance tasks easy by responding to critical aspects of any government or industry IT security regulation and providing proof of compliance for audit purposes.

PAM provides a central management console that enables quick, streamlined management of all users across multiple or disparate systems, including and especially hybrid cloud environments. PAM as a single point of control gains time, efficiency, and oversight over all privileged users active in a hybrid cloud environment. A PAM solution pattern generates an unalterable audit trail for any privileged operation, so the IT security team can track, view, or replay the actions of any privileged user.

IAM creates audit logs for all successful and unsuccessful authentication attempts by users. Logs are also created for privileged access to a cloud platform, services, and resources. Cloud providers use **security information and event management** (**SIEM**) to monitor and analyze logs and report any security incidents. They share the subset of their monitoring and audit reports with their clients, based on mutual contract agreements.

Known uses

The following table provides examples of each cloud IAM pattern and how it is implemented by different cloud service providers:

IAM pattern	AWS	Azure	Google Cloud	IBM Cloud
Self-registration	AWS provides sign-up of free accounts and a consolidated billing option to register for the cloud services: `aws.amazon.com/free`	Azure free account and "pay-as-you-go" sign-up: `https://azure.microsoft.com/en-in/free/`	GCP free trial registration link: `https://console.cloud.google.com/freetrial`	IBM Cloud free account: `https://www.ibm.com/in-en/cloud/free`
Federation	AWS supports federating enterprise IDs of a workforce as part of AWS accounts and business applications: `https://aws.amazon.com/identity/federation/`	Federation with **Azure Active Directory** (**Azure AD**): `https://docs.microsoft.com/en-us/azure/active-directory/hybrid/whatis-fed`	Best practices for federating Google Cloud with an external identity provider: `https://cloud.google.com/architecture/identity/best-practices-for-federating`	*IBM Cloud SAML Federation Guide*: `https://www.ibm.com/cloud/blog/ibm-cloud-saml-federation-guide`
Cloud identity	AWS Directory Service – a managed Microsoft active directory in AWS: `https://aws.amazon.com/directoryservice/`	Azure AD: `https://azure.microsoft.com/en-in/services/active-directory`	Google Cloud Identity: `https://cloud.google.com/identity`	IBM Cloud Identity Solution: `https://www.ibm.com/in-en/security/cloud-identity`
Service accounts	An AWS IAM user can represent a person or an application that uses its credentials to make AWS requests: `https://docs.aws.amazon.com/IAM/latest/UserGuide/id_users.html`	Azure AD service accounts: `https://docs.microsoft.com/en-us/azure/active-directory/fundamentals/service-accounts-governing-azure`	Google Cloud service accounts: `https://cloud.google.com/iam/docs/service-accounts`	Creating and working with service IDs: `https://cloud.ibm.com/docs/account?topic=account-serviceids`

IAM pattern	AWS	Azure	Google Cloud	IBM CLOUD
User de-provisioning	Managing IAM users: `https://docs.aws.amazon.com/IAM/latest/UserGuide/id_users_manage.html`	Deleting a user: `https://docs.microsoft.com/en-us/azure/active-directory/fundamentals/add-users-azure-active-directory#delete-a-user`	Revoking access to Google Cloud Platform: `https://cloud.google.com/security/data-loss-prevention/revoking-user-access`	Removing users from an account: `https://cloud.ibm.com/docs/account?topic=account-remove`
Logging in with user ID and credentials	Signing in to AWS: `https://aws.amazon.com/`	Signing in to the Azure portal: `https://azure.microsoft.com/en-in/account/`	Google Cloud sign-in: `https://console.cloud.google.com/home/dashboard`	IBM Cloud login: `https://cloud.ibm.com/login`
Application access key	Managing access keys for AWS users: `https://docs.aws.amazon.com/IAM/latest/UserGuide/id_credentials_access-keys.html`	Passwordless authentication: `https://docs.microsoft.com/en-us/azure/active-directory/authentication/concept-authentication-passwordless`	Using API keys: `https://cloud.google.com/docs/authentication/api-keys`	Managing user API keys: `https://cloud.ibm.com/docs/account?topic=account-userapikey`
SSO	AWS SSO: `https://docs.aws.amazon.com/singlesignon/latest/userguide/what-is.html`	Azure Active Directory SSO: `https://azure.microsoft.com/en-us/services/active-directory/sso/`	Google Cloud SSO: `https://cloud.google.com/architecture/identity/single-sign-on`	Logging in with federated identity: `https://cloud.ibm.com/docs/account?topic=account-federated_id`
SSH keys	Amazon key pairs: `https://docs.aws.amazon.com/AWSEC2/latest/UserGuide/ec2-key-pairs.html`	Create and use SSH keys for Linux VMs: `https://docs.microsoft.com/en-us/azure/virtual-machines/linux/mac-create-ssh-keys`	SSH with security keys on Compute Engine: `https://cloud.google.com/compute/docs/tutorials/ssh-with-sk`	Adding an SSH key: `https://cloud.ibm.com/docs/ssh-keys?topic=ssh-keys-adding-an-ssh-key`

IAM pattern	AWS	Azure	Google Cloud	IBM CLOUD
MFA	AWS MFA: `https://aws.amazon.com/iam/features/mfa/`	Azure MFA: `https://docs.microsoft.com/en-us/azure/active-directory/authentication/concept-mfa-howitworks`	Enforce MFA: `https://cloud.google.com/identity/solutions/enforce-mfa`	Enabling MFA for your account: `https://cloud.ibm.com/docs/account?topic=account-enablemfa`
SLO	AWS logout: `https://docs.aws.amazon.com/singlesignon/latest/PortalAPIReference/API_Logout.html`	Single Sign-Out: `https://docs.microsoft.com/en-us/azure/active-directory/develop/single-sign-out-saml-protocol`	Single sign-off supported through Cloud Identity: `https://cloud.google.com/architecture/identity/single-sign-on`	IBM Cloud logout: `https://iam.cloud.ibm.com/identity/logout`
Physical authentication	Physical access: `https://aws.amazon.com/compliance/data-center/controls/#Physical_Access`	Azure facilities, premises, and physical security: `https://docs.microsoft.com/en-us/azure/security/fundamentals/physical-security`	Google infrastructure security design: `https://cloud.google.com/security/infrastructure/design`	Physical security architecture: `https://www.ibm.com/cloud/architecture/architectures/physical-security-arch/`
Access control	AWS IAM: `https://aws.amazon.com/iam/`	Azure RBAC documentation: `https://docs.microsoft.com/en-us/azure/role-based-access-control/`	Google Cloud IAM: `https://cloud.google.com/iam`	IBM Cloud IAM: `https://cloud.ibm.com/docs/account?topic=account-iamoverview`
Identity governance and administration	AWS security audit guidelines: `https://docs.aws.amazon.com/IAM/latest/UserGuide/id_credentials_getting-report.html` Access Analyzer: `https://docs.aws.amazon.com/IAM/latest/UserGuide/what-is-access-analyzer.html`	Azure security logging and auditing: `https://docs.microsoft.com/en-us/azure/security/fundamentals/log-audit`	Audit logging: `https://cloud.google.com/iam/docs/audit-logging` Policy Intelligence tools: `https://cloud.google.com/iam/docs/policy-intelligence-tools`	Auditing access to resources: `https://cloud.ibm.com/docs/account?topic=account-access-report`

Table 3.1 – IAM patterns and known uses

Related patterns

IAM patterns for applications will be discussed in the next chapter.

Summary

In this chapter, we discussed IAM, which is the core component for managing identity across a hybrid cloud. We captured the important patterns for operationalizing IAM for users of a cloud platform. We included the key patterns for user management, authentication, authorization, governance, and administration.

In the next chapter, we shall discuss how to tackle the same problems and solution patterns to manage identity in cloud applications.

4

Implementing Identity and Access Management for Cloud Applications

In this chapter, you will learn how to add authentication and manage access to applications deployed on the cloud. This chapter will discuss the patterns to enhance web and mobile apps with **identity and access management (IAM)** security capabilities.

The following topics will be covered in this chapter:

- Authentication pattern for cloud application users
- Service-to-service authentication
- Cloud application authorization patterns

The following diagram illustrates application IAM patterns:

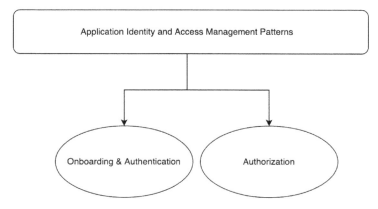

Figure 4.1 – Application IAM patterns

As shown in the preceding diagram, the patterns cover user onboarding, authentication and authorization patterns for web or mobile, and **application programming interface (API)** and backend applications.

Authentication pattern for cloud application users

Let's get started!

Problem

How to manage the onboarding and authenticating of users for a cloud-deployed application.

Context

Modern cloud applications are focused on providing a personalized experience to their users. The application users can be the enterprise's employees, customers, and partners. To provide a personalized experience for these users, it is important to identify the users and related attributes. A cloud application may be typically developed and deployed following cloud-native technologies, as a set of microservices, leveraging resources and services from multiple clouds. The problem to be solved is how to authenticate the user to the cloud-deployed applications and establish trusted access.

For applications used by employees, most enterprises want to use their existing **identity provider (IdP)** solutions so that their employees can sign on with their enterprise **user identifier (UID)** and credentials. Similarly, customers and partners would also like to be onboarded to cloud applications using their own company UID and credentials.

End users such as customers may wish to bring in their own identity (**Bring Your Own Identity**, or **BYOI**) or use their social IDs such as Facebook, Google, or GitHub IDs to access the cloud application. Using their existing IDs makes it easy for users to log in to the applications. In certain other cases, users may need to be onboarded **just in time (JIT)**, whereby their ID and credentials are created when they register or sign up for the first time. This is the scenario for enterprises that do not have a known list of their application users or customers, such as applications attracting new customers not previously known to the enterprise. They want to enable their customers to *join* the application or sign up through a self-service portal.

Another dimension to this problem context is that users of the application like to do progressive authentication—anonymous user get to do progressive authentication. In this model, the application allows the user to browse through unprotected websites without having to log in. Once they sign up or identity themselves, then the user expects personalized experiences based on their attributes or details shared. An unprotected or informational website may not need to identify the user and does not need authentication.

Modern cloud applications allow users to experience parts of the application without having to sign up, and once they like the experience and they need more of it, the application encourages them to sign up. This is a transition from an anonymous user identity to establishing an identity through progressive authentication. For example, users can navigate or add items to a shopping cart through an app without signing up. When they need to check out items or buy items, they need to register and authenticate to make payments.

Modern applications designed in a cloud-native programming model are composed of several microservices running on different clouds. So, the authentication mechanism must support the authentication of users for applications written in multiple languages and running on different clouds—public or private.

Solution

The cloud application's user type or profile determines the type of authentication method that needs to be used.

Application IAM solutions typically support standards-based integration with several IdPs as well as custom protocols for authentication. The authentication patterns supported by an application IAM solution are shown in *Figure 4.2*. The entire application IAM solution is available as a service from the cloud and is an integral service in most **cloud service provider** (**CSP**) catalogs.

As shown in the following diagram, the application IAM supports the onboarding and authentication of the different types of users, as discussed in the problem context, through different protocols:

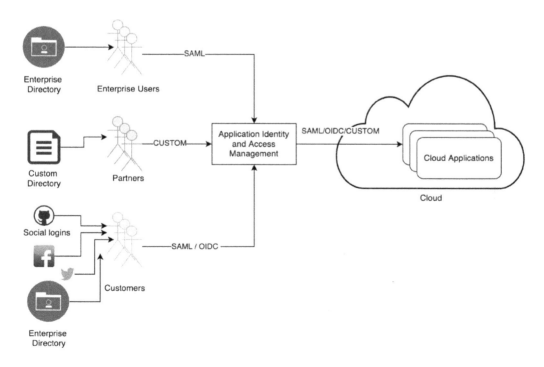

Figure 4.2 – Application onboarding and authentication pattern

Employers, customers, or partners that want to leverage their existing user directory and identity infrastructure to authenticate their users perform integrations with the application that are typically done based on standards-based protocols such as **Security Assertion Markup Language** (**SAML**) and **OpenID Connect** (**OIDC**). These standards help avoid duplication of identities, credentials, and other attributes. Depending on these attributes, users can be put into different groups such as anonymous groups that can have certain rights. Once a user signs up (creates an account, with some amount of validation of the user, even if only self-validation), that user is added to the *KnownCustomers* group, which has the rights to do things such as make purchases. And it's also good to note that users can be, and often are, in multiple groups, such as *KnownCustomer* and *StoreEmployee* groups, concurrently. Those groups have different rights, of course.

For users that want to do a JIT sign-up with the app, the requirement is typically fulfilled with a Cloud Directory capability. The Cloud Directory pattern, discussed in *Chapter 1*, may be leveraged for application authentication. Cloud Directory lets the mobile and web applications delegate the management of the users' details and enable quick sign-up and sign-in. Cloud Directory provides a user registry for applications that scales seamlessly and provides basic authentication such as requiring the user's email address, mobile number, and password.

Most IdPs support standard protocols such as SAML and OIDC, but a few enterprises still have legacy proprietary identity systems. Wherever standards are missing, users need to integrate with their identity system using custom authentication protocols. In some cases, user details are stored in a file or a database. Integrating these kinds of custom IdPs for application authentication can be painful and, in fact, storing user details in this way creates other security exposures, such as the fact that the file could, itself, be stolen. If possible, the opportunity should be taken to migrate such approaches to a more modern IAM.

Let's look at the SAML and OIDC protocols in detail next.

SAML

SAML is an open standard for exchanging security information between entities. The building blocks and participants of SAML exchange identity, authentication, attribute, and authorization information. SAML supports web **single sign-on** (**SSO**) and identity federation use cases for cloud applications.

As shown in the following diagram, the SAML interaction starts with a user trying to log in to a cloud application. SAML exchanges take place between the cloud application, IdP, and the user accessing the application through a web browser. A login session or a security context is created to evaluate accessing resources managed by that cloud application. This SAML flow, which is typically called a **service provider** (**SP**)-initiated flow, is initiated by the cloud application that is also referred to as the SP. In *step 2*, the cloud application redirects the user to a SAML IdP. A trust relationship is pre-established between the cloud application and the IdP. The IdP redirects the user to a login page. The user then enters the UID/credentials, which are verified against the user identity store. The verification response is provided back to the IdP, which in turn provides a SAML response to the cloud application:

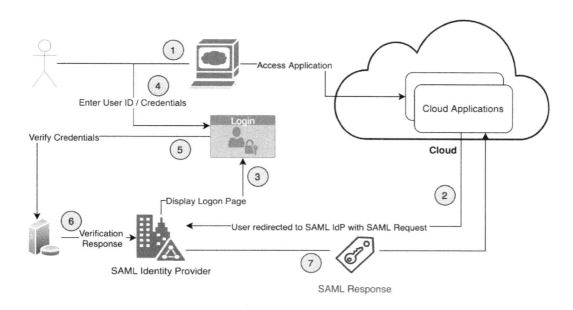

Figure 4.3 – SAML protocol

Based on the response from the IdP, the cloud application can make an *authorization* decision based on the assertions in the SAML response. The SAML response does not contain any user credential information. The SAML specification also defines a structure, schema, and protocol for the exchange of these messages between the parties. Communication between the SP and IdP is encrypted and encoded for security purposes. Refer to the SAML site for more details on the standard (http://docs.oasis-open.org/security/saml/Post2.0/sstc-saml-tech-overview-2.0.html).

OIDC

There will be a population of application users that want to bring in their identity (BYOI) or social logins to authenticate themselves for access to the application. This model is again managed through standards-based integration. The most frequently used standard is OIDC. Enterprise **identity management (IdM)** is also transitioning from SAML to OIDC as a preferred protocol for integrating with cloud applications.

OIDC is a lightweight identity layer built on top of the **Open Authorization (OAuth)** protocol. OIDC supports cloud applications to verify the identity of the user based on the authentication performed by an authorization server. This authentication information is shared in an interoperable and **REpresentational State Transfer (REST)**-like manner: friendly and secure.

As shown in the following diagram, OIDC allows the cloud application user's authentication to be done with the help of an IdP:

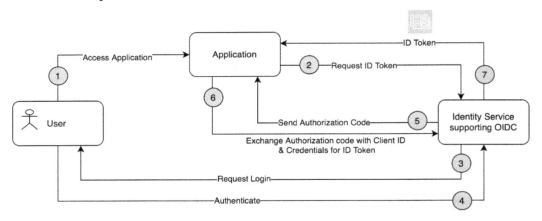

Figure 4.4 – OIDC protocol

The user logs in to any cloud application that supports OpenID authentication by using their accounts. OpenID supports SSO by allowing single credentials for authenticating to multiple cloud applications. The application redirects the user to an identity service supporting the OIDC protocol. The **identity SP (IDSP)** redirects the user to the login page for the user to enter their user details and credentials. The IdP stores the user details, and the applications create accounts for their use. The identity service returns an authorization code to the requestor. In this authorization code flow pattern, the application exchanges the authorization code for a token. This exchange requires a client secret that has to be available with the cloud application. Typically, a server-side application requires an end user; however, it relies on interaction with the end user's web browser, which redirects the user and then receives an authorization code. On successful authentication, the IdP returns the user profile claims as a token against the authorization code. This identity token contains authentication information. OIDC recommends **JavaScript Object Notation (JSON)** as the data format for exchanging information in a language-neutral, text-based, and lightweight manner. The tokens exchanged are implemented as **JSON Web Tokens (JWTs) (Request for Comments (RFC)** *7519*) that clients receive after successful user authentication.

The claims could vary, depending on the backend authentication service. JWT is a JSON format useful for exchanging information between parties. It contains most claims found within access tokens, but also includes user identity claims, such as name, email, locale, and picture. Identity tokens are used to understand who the user is. User identity claims come from IdPs and depend on how the application requires users to authenticate. Both tokens are cryptographically signed with tenant-specific keys to prevent tampering and contain a set of claims that help developers to customize application behavior and make business logic decisions. Please refer to the links provided in the *References* section for details on the OIDC standard and protocol, as well as details on JWT.

Known uses

We will look at some known uses here:

- **AWS Cognito** (`https://aws.amazon.com/cognito/`) is a secure and scalable identity store as well as supporting users signing in to cloud applications using social and enterprise identities. Amazon Cognito supports application identity integration with social IdPs such as Apple, Google, Facebook, and Amazon, and enterprise IdPs leveraging SAML and OIDC patterns.

- **Azure Active Directory (Azure AD)** (`https://docs.microsoft.com/en-us/azure/active-directory/external-identities/identity-providers`) is a multi-tenant, cloud-based IAM service that supports IdPs for external identities, including social IdPs such as Google and Facebook, and setting up federation with any external IdP that supports the SAML or **Web Services Federation (WS-Fed)** protocols.

- **Google Identity Services (GIS)** (`https://developers.google.com/identity`) enables users to sign in to apps and authorize apps to use Google services. Android and web authentication API flows support the custom login widget for sign-in and sign-up options. GIS APIs conform to the OIDC specification for both authentications.

- **IBM Cloud App ID** (`https://www.ibm.com/cloud/app-id`) allows easy adding of authentication, secure backends, and APIs, and managing user-specific data for mobile and web applications. App ID allows you to enhance applications with advanced security capabilities, such as **multi-factor authentication (MFA)**, SSO based on standards, and user-defined password policies. App ID also provides a scalable user registry for users to manage their own accounts.

Service-to-service authentication

Let's get started with service-to-service authentication!

Problem

How to onboard and authenticate services for a cloud-deployed application.

Context

Apart from users, systems and services interact with cloud resources and cloud applications to deliver functionality. There is a need for a standards-based approach to be able to implement effective and efficient service access to applications in the cloud.

Traditionally, security is enforced at the edge layer for a monolithic application, and subsequent interactions are considered secure. This is like an entry check at the gate of a mansion, and then no authentication is required to enter different rooms of a building. But in cloud-based applications, there are multiple microservices, and they need to communicate with each other to deliver the function. Authentication is required at each service entry point. This is like an identity check for entering each room in the mansion. These services are likely to communicate with each other using either asynchronous or synchronous means. Many of these services may be private or require explicit authentication for access. One service calls another receiving service, using its endpoint **Uniform Resource Locator (URL)**. Application IAM is required at the boundary of each of these services to establish a service identity. For the receiving service, the user identity on behalf of whom the service is invoked also needs to be considered for determining the set of permissions needed to perform the work.

Another example is where a client, such as a mobile application, directly invokes a cloud service or needs access to a protected resource on the cloud, such as a backend API. The actions or the backend APIs invoked could be different based on the user identity. If we consider this as north-south direction service interaction, there is also a need to secure east-west interactions between the services on the same cloud or on different clouds. Please see the following *Solution* section for details on patterns applicable for each case.

Solution

A cloud application can have its own identity and credential (application access key or public key certificate) and use that identity to authenticate and access any cloud resource. The service access to services such as storage or runtimes is subject to access-control rules specified based on those resources. In a cloud-native programming model, an application is formed of many services. Each service can also have its own identity. Please read the concept of **service ID (SID)** or service account, discussed in *Chapter 3*, for details. This identity is used to mutually authenticate with an invoking application or with other cloud services. With the use of application identity and credentials passed as a token, cloud applications can uniquely identify other applications or services. By using an access token, protected resources are provided access only after validation of the token to establish it is originating from a trusted, authenticated source.

You can see an illustration of the process in the following diagram:

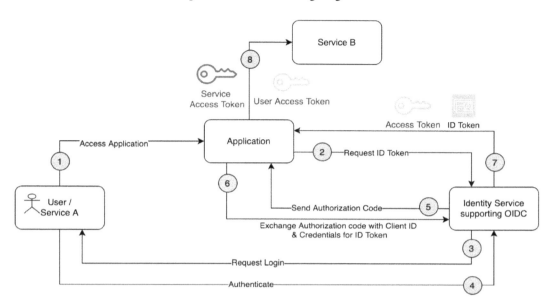

Figure 4.5 – Service-to-service authentication

For several reasons, one application needs to communicate with another service or application without any user intervention. For example, as shown in the preceding diagram, a non-interactive or backend application needs to access another application to perform its work. This application could have requirements to access another set of cloud services to fulfill its tasks. For instance, if service *A* needs to access service *B*, it needs to propagate its own service identity as well as additional information such as the invoker or the user identity. The important aspect to consider is that the requests are exchanged on behalf of the application, not on behalf of an end user. So, the first check is to ensure that the service or application is authenticated and authorized to access the target service or application. But depending on the application design, access control for service *B* could be based on the propagated user identity and not the invoking service's identity. The identity is propagated in the form of tokens, as discussed in the pattern for user authentication for cloud applications. There are multiple patterns for managing authentication and authorization for service-to-service interactions, depending on the use case.

Client application accessing a protected resource on the cloud

In this pattern, as shown in the following diagram, the user accesses the cloud resource through a client-side application such as a mobile application. The protected resource or service is invoked on behalf of the end user. The user identity token needs to be passed to make a successful invocation. The cloud service or protected resource, upon receiving the tokens, can validate them and action the request accordingly. As discussed in the preceding pattern, the receiving cloud service may need to interact with other services to deliver the functionality. In this case, the tokens are propagated downstream for user authentication and authorization:

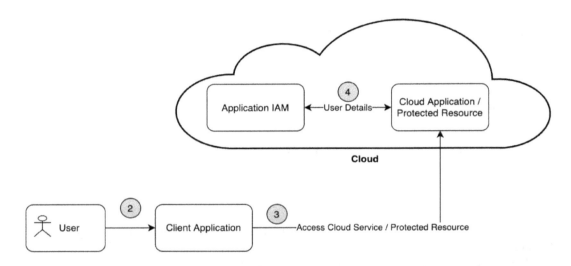

Figure 4.6 – Client application directly accessing cloud resource

Application accessing a protected resource on the cloud through a gateway

The gateway pattern, as shown in the following diagram, is very similar to the pattern of the client application directly accessing the protected resource on the cloud. The variation in this pattern is that security is enforced at the edge or gateway. This assumes that subsequent layers are considered secure:

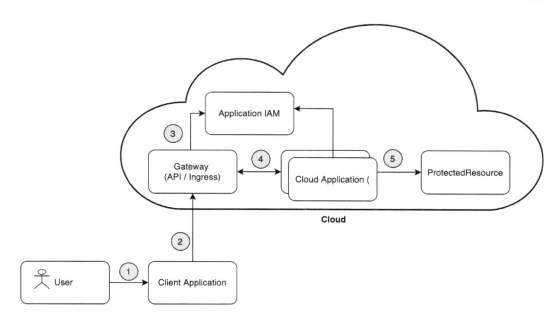

Figure 4.7 – Managing authentication and authorization at the gateway

Once the identity is authenticated at the gateway level, subsequent services need not enforce checks.

Application accessing a protected resource on the cloud through another service

As shown in the following diagram, in the pattern where trust cannot be established with client applications such as browser applications, we can take the approach of routing access to the protected resource through a server-side service:

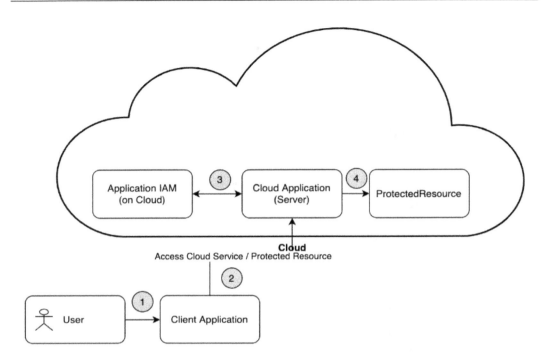

Figure 4.8 – Managing authentication and authorization through a server-side service

The variation from the preceding patterns of clients directly or through a gateway is that the identity tokens remain on the server side and are not exposed outside. In this pattern, a user requests a backend resource through the client application. The client application interacts with the backend service or server application. The backend service then invokes the service to access the protected service or resource on the user's behalf. All the interactions may not involve an end user. Based on mutual trust between the microservices, one service can get results from another service and pass back the information to the client service, which provides it to the user.

Service mesh pattern

For certain scenarios, a central gateway model fails to address the security need for agility and evolution of a microservices model that needs protection for inter-service interactions' cloud environment (east-west communications) as well. In this service mesh pattern, there is a control plane that manages the overall topology and configuration. The data plane of the service mesh manages communication between the services.

As shown in the following diagram, the service mesh pattern is non-intrusive or does not introduce any new functionality in the service or application:

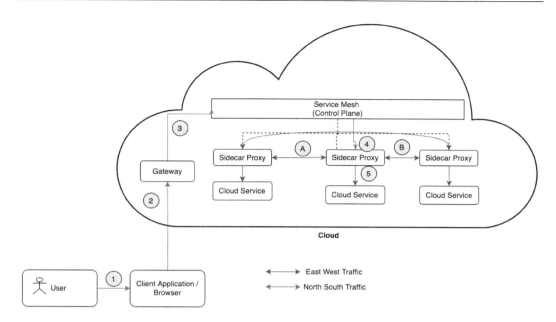

Figure 4.9 – Service mesh pattern

With this architecture, the service-to-service communication policy and governance are managed by the control plane as opposed to the individual services. The security functionality in the service mesh pattern is achieved by a set of sidecar proxies. Sidecar proxies sit in between the client and target service, check for security, and then forward to the target service endpoint. Similarly, responses are verified and security is checked by the proxy before returning to the invoker.

The service mesh pattern can be extended across multiple cloud and on-premise environments, making it a good choice for implementing micro-perimeter-level authentication and authorization, leveraging a central or federated control plane.

Some of the merits of adopting this pattern include the following:

- With the service mesh pattern, there is no need to duplicate the functionality or repeat the code in each service for enforcing authentication and authorization

- This is aligned with the zero-trust security model that goes beyond perimeter-based protection and enforces security for each microservice (micro-perimeter level) even inside the trusted zone

- The service mesh pattern can also enforce mutual trust through **Mutual Transport Layer Security (mTLS)** for inter-service communication

- The service mesh can help protect against service impersonation or unauthorized access as only authenticated services can communicate through the mesh

- Any risk of insider threat related to data exfiltration or **man-in-the-middle (MITM)** attacks between clients and protected resources can also be eliminated with the service mesh pattern

Known uses

We will look at some known uses here:

- AWS Cognito, Azure AD, GIS, and IBM App ID services support patterns accessing a protected resource on the cloud through a gateway pattern as an API gateway solution integrated with the application IAM service.

- Similarly, access to microservices or backend APIs running on Kubernetes clusters may be protected at the ingress level with authentication. The ingress controller provides a mechanism where you can define a policy that will be enforced to delegate all needed authorization and authentication to the ingress controller of the applications.

- Istio, Linkerd, AWS App Mesh, and Consul Connect are some of the top service mesh tools that support communication between microservices in a Kubernetes environment. They can be leveraged to build service mesh authentication and authorization patterns.

Cloud application authorization patterns

Let's get started with cloud application authorization patterns!

Problem

How to manage access to different application functionality for users or services.

Context

Authorization determines what a user or a service is permitted or not permitted to do inside the application. Once a user or service identity is established, there is a need to establish which actions they can perform with the application.

Solution

Authorization is a method of allowing or denying access to a particular resource depending on an authenticated user's or service's entitlements. The authorization could work at two levels, as follows:

- **Coarse-grained**—High-level and overarching entitlements defined as **create, read, update, delete (CRUD)** sorts of permissions at the service or protected resource level

- **Fine-grained**—Detailed, explicit, and specific entitlements to protected resources based on additional factors such as time, location, department, role, and other attributes

Authorization in a cloud environment is attained by either access-control policies or access right delegations. The authorization process decides which user or service is allowed to perform certain operations on the system or application.

Authorization rights can be managed centrally, which keeps sensitive information in one place. This model has reduced management and security costs for keeping the information secure. Alternatively, it could be handled by each third-party application in a distributed style. This has a risk that the user's private information can be accessed by multiple applications. Such permissions are unsafe since the privacy of the user is involved.

A cloud network contains different SP environments in which a single user is able to access different kinds of services at the same time while each service is from a different SP and with different security levels. Sometimes, authorization rights are given by third-party vendors, and these third-party applications are authorized to access certain private information. Such permissions are unsafe since the privacy of the user is involved. Authorization in a cloud environment is attained by either access-control policies or access-right delegations. The CSP defines and implements access-control policies such that the resources and services are accessed only by authorized users. Centralized access-control mechanisms are advantageous to organizations in securing sensitive information and reducing several management and security tasks.

Cloud applications implement one or a combination of multiple of the following access-control patterns to address the authorization requirements.

Mandatory access control

Mandatory access control (MAC) takes a confidentiality-first approach. It limits access based on the sensitivity of the information and the need to know from a user's perspective.

You can see a depiction of this in the following diagram:

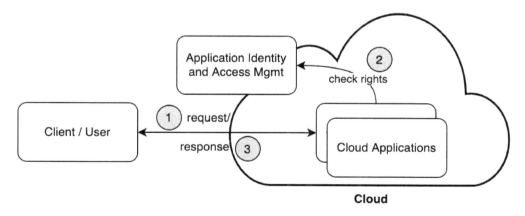

Figure 4.10 – MAC

Security levels or labels are assigned to the resources based on sensitivity criteria such as *internal*, *external*, *internal restricted*, *restricted*, and *confidential*. Users or systems can access information based on the security level or labels that they are entitled to. The labels or levels may be grouped as categories. Only if the user is assigned access to a specific category can they access that protected resource.

Discretionary access control

A **discretionary access control** (**DAC**) pattern is leveraged for implementing authorization or controls where MAC is lacking. In this model, access is determined by discretion.

You can see a depiction of this in the following diagram:

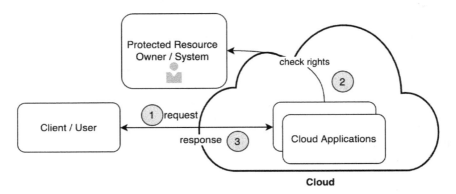

Figure 4.11 – DAC

Controls are identified by the group to which the subject belongs and whether the permissions to access the protected resource are acquired by the subject directly or indirectly. In this case, access is determined based on the information of the subjects or objects, their attributes, and existing governance rules.

Role-based access control

In **role-based access control** (**RBAC**), users of an application are assigned a set of permissions or actions on specific resources. These permissions are grouped together as a **role**. Users are then assigned to these roles, such as *employee*, *end user*, *vendor*, *administrator*, or *privileged* user.

You can see a depiction of this in the following diagram:

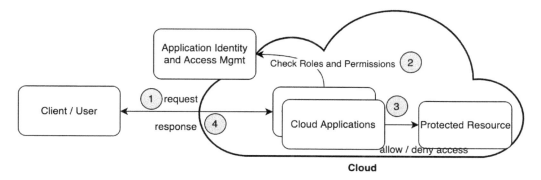

Figure 4.12 – RBAC

The RBAC model ensures that the user has the right permissions and access to the resources to perform their job. The roles could be based on several factors—business attributes such as the employee's job responsibility, as well as technical attributes such as specific tasks requiring access to special resources such as a database or record. The roles are updated as employees switch roles or move jobs. With a predefined set of roles, RBAC is one of the easiest patterns to implement in an authorization model by assigning users to roles. Roles can also be discovered or mined based on the analysis of the workforce and their existing permissions or access needs. RBAC requires that there is a periodic audit for validating whether the user is in the same role and whether continued access is needed.

Attribute-based access control

Attribute-based access control (**ABAC**) is a logical access model that combines an **access-control list** (**ACL**) and RBAC. Access is provided based on the evaluation of specific attributes or characteristics rather than roles.

You can see a depiction of this in the following diagram:

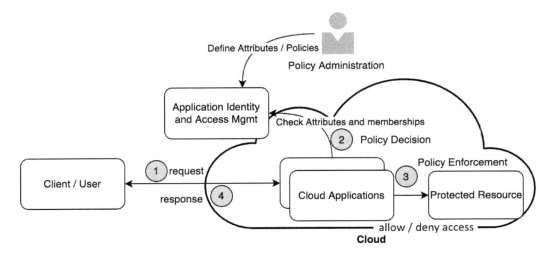

Figure 4.13 – ABAC

ABAC is more context-aware—that is, access to a specific database or functionality in an application is determined based on an evaluation of the attributes of the user and the criticality of the information. For example, a sales user trying to update a critical lead in a sales management application beyond business hours logged on to the application from a different network or time zone may be denied access based on an evaluation of the risk along with context and attributes.

Entitlement/task-based access control

Task-based access control (**TBAC**) is similar to ABAC, whereby access to protected resources is provided only in the context of completing a specific task.

You can see a depiction of this in the following diagram:

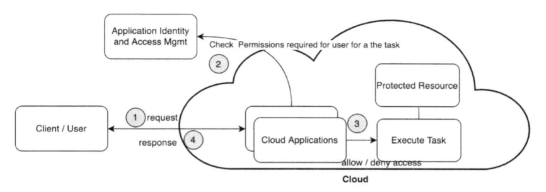

Figure 4.14 – TBAC

Typically, this is used for service accounts or system IDs where tasks or workflows are executed at a certain time of the day. For example, access to sensitive health information for a healthcare worker or accessing banking statement details by a teller persona is based on the need for them to complete the task at hand.

Access delegation model (OAuth)

In this authorization model, the resource owner gives permission to a service or an application to access a protected resource. OAuth 2.0 is the security standard that defines the protocol for this model. In this model, the requestor service gets a key or token to access the protected resource without sharing any UID or credentials. This is a delegation model where the validity of the access token can also be set to a specific time interval.

You can see a representation of this model in the following diagram:

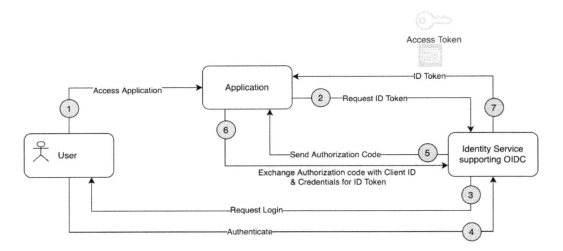

Figure 4.15 – Access delegation pattern

In this model, a **relying party** (**RP**) is a resource provider or an application that requires end-user verification. An access token represents authorization to the bearer. It contains claims (JSON properties) that describe who the token was issued by, who it was issued for, the intended audience, the token expiration timestamp, which authentication method was used, the authorization scope, and more. Access tokens are used to understand what the user is authorized for. The application can implement these tokens with highly advanced logic and customized access-control solutions. Application developers can take advantage of the **Software Development Kit** (**SDK**) if one is supplied by the application IAM service to implement the logic in their preferred language and platform. An alternative approach is to take a language-agnostic approach to interact with application IAM through REST APIs.

Known uses

We will look at some of the known uses here:

- Amazon Cognito (https://aws.amazon.com/cognito/) provides access control to resources from the application. We can define roles and map users to different roles so that the application can access only the resources that are authorized for each user or service.

- Azure AD (`https://docs.microsoft.com/en-us/azure/active-directory/external-identities/authentication-conditional-access`) supports a conditional access pattern. Organizations can enforce multiple conditional access policies for their external users, which can be enforced at the tenant, application, or individual user level in the same way that they're enabled for full-time employees and members of the organization. This solution leverages OAuth 2.0 for authorization.

- The authorization feature with GIS provides an SDK or library that can be leveraged by the application for both authentication and authorization.

- Applications can leverage the IBM Cloud App ID (`https://cloud.ibm.com/docs/appid?topic=appid-app`) capability to store user preferences and data to provide a personalized experience. App ID leverages the OAuth 2.0 client credentials flow to protect communication. After an app registers with App ID, the app obtains a client ID and secret. With this information, the app can request an access token from App ID and be authorized to access a protected resource or API. In the application identity and authorization flow, the application is granted only an access token.

Summary

In this chapter, we learned about the various patterns that we can use to add IAM capabilities for cloud applications. This addresses how to meet the authentication and authorization needs of an application that leverages resources and services from the cloud. We also looked at how to deal with service-to-service authentication and authorization patterns. Building these modules in a standards-based model helps accelerate the implementation. In this regard, we learned the specifics of standards such as SAML and OIDC.

In the next chapter, we will learn about the patterns on how to secure the hybrid cloud infrastructure on top of which these applications run.

References

To learn more about the topics covered in this chapter, you can visit the following links:

- Amazon Cognito—`https://aws.amazon.com/cognito/`

- Azure AD—`https://docs.microsoft.com/en-us/azure/active-directory/`

- GIS—`https://developers.google.com/identity`

- IBM App ID—`https://www.ibm.com/cloud/app-id`

- Istio service mesh—`https://istio.io/latest/about/service-mesh/`

- Linkerd service mesh—`https://linkerd.io/`

- Consul service mesh—`https://www.consul.io/docs/connect`
- AWS App Mesh—`https://aws.amazon.com/app-mesh/`
- SAML—`https://en.wikipedia.org/wiki/Security_Assertion_Markup_Language`, `http://docs.oasis-open.org/security/saml/Post2.0/sstc-saml-tech-overview-2.0.html`
- OpenID—`https://openid.net/connect/`
- OAuth—`https://oauth.net/2/`
- JWT—`https://jwt.io/`
- JSON—`https://www.json.org/json-en.html`
- DAC—`https://en.wikipedia.org/wiki/Discretionary_access_control`
- MAC—`https://en.wikipedia.org/wiki/Mandatory_access_control`

Part 3:
Infrastructure Security Patterns

Enterprises can take advantage of the variety of compute options that are now available in hybrid multi-cloud environments, namely: bare-metal servers, VMs, containers, and serverless, as well as specialized infrastructure to develop and run their software. This chapter provides the context to the type of attacks that can happen against different compute types of the modern hybrid cloud infrastructure and how to protect them. It is also important to know how to isolate network traffic based on purpose and establish secure connectivity across these compute environments. In this part, we will discuss the patterns on how to secure the hybrid cloud compute and network, which are the most essential components of hybrid cloud infrastructure.

This part comprises the following chapters:

- *Chapter 5, How to Secure Compute Infrastructure?*
- *Chapter 6, Implementing Network Protection, Isolation, and Secure Connectivity*

5

How to Secure
Compute Infrastructure

In this chapter, we will learn about the patterns that can be leveraged to secure a hybrid cloud compute infrastructure. A modern hybrid cloud infrastructure consists of the following compute types:

- Bare-metal servers
- **Virtual machines (VMs)**
- Containers
- Serverless

Depending on the type of compute, the pattern for securing them also varies. The following diagram shows the different protection patterns for compute that will be discussed in this chapter:

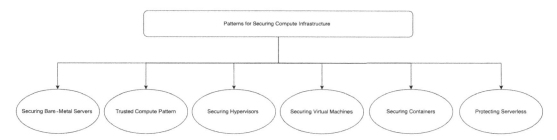

Figure 5.1 – Patterns for securing cloud compute infrastructure

In the shared responsibility model for cloud security, the roles and responsibilities between the cloud provider and consumer change based on the compute type consumed from the cloud. We will discuss patterns to provide isolation to varying degrees and security for bare-metal servers, VMs, containers, and serverless compute types.

We will cover the following main topics in this chapter:

- Securing physical (bare-metal) servers
- Trusted compute patterns
- Securing hypervisors
- Protecting VMs
- Securing containers
- Securing serverless implementations

Securing physical (bare-metal) servers

Let's get started!

Problem

How to secure and protect bare-metal servers.

Context

The foundation infrastructure is made up of bare-metal or physical servers. Bare-metal servers are also referred to as **dedicated servers** and provide maximum performance by delivering single tenancy. As shown in the diagram that follows, the security design needs to cover the different layers that make up the server. These layers include the physical hardware and the host operating system forming the bottom of the stack. The next layer is formed of the binaries and libraries that are leveraged by the operating system and hosted applications:

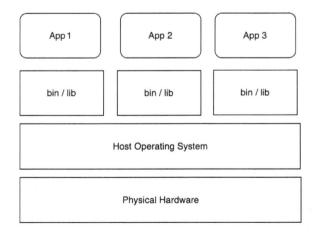

Figure 5.2 – Bare-metal server

The bare-metal server option provides direct root access to server resources and the consumer has the opportunity to customize the environment as per their needs. While there is flexibility with a bare-metal server, it increases the security risks. Server security should address the protection of critical components and services from exploits through the network as well as through native attacks. For bare-metal servers, the cloud provider takes essentially no responsibility for logical security, though it typically does take responsibility for the physical security of the servers.

Solution

The primary solution pattern for securing bare-metal servers is called **server hardening**. The hardening of a server is the process by which the security posture is improved by limiting the server's exposure to threats and vulnerabilities. Hardening looks to reduce the attack surface and vectors. In general, a single-function system is likely to be more secure than a multifunction one. The larger the vulnerability surface, the more the server is subject to threats. Hardening aims to reduce the number of attack vectors, while threat prevention and vulnerability detection measure aims, respectively, to close entry points for attack vectors and to detect them should they exist.

You can see a visual representation of this in the following diagram:

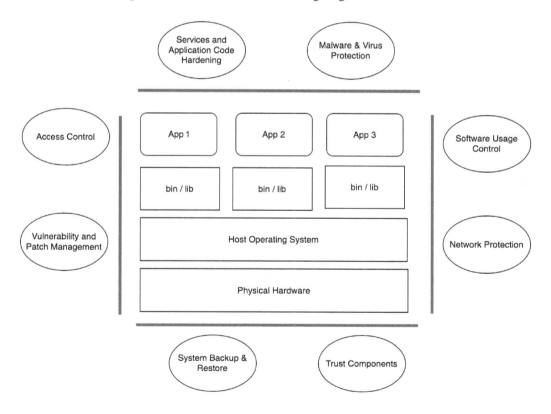

Figure 5.3 – Bare-metal protection

Let us look at the process of hardening in more detail, as follows:

1.   The first step of hardening is to review and remove unused or unwanted users from the system that have privileged access.

2.   The next step is to disable unwanted services on the system, as well as removing unnecessary software. This will reduce the attack surface and vulnerabilities in the *bin/lib* layer.

3.   The next logical step is to ensure the operating systems and software are free from vulnerabilities. This step of hardening is done by setting the proper configuration and patching. Patching is the process of fixing or handling security vulnerabilities that can pose a risk to a business's data and information. Details on vulnerability management are discussed in *Chapter 11*.

4.   Protecting the server through network hardening is required to ensure only the right connectivity to the outside world and the internet is provided to the server. For management functions, a separate private network needs to be planned. Details on network protection, isolation, and management are discussed in the next chapter.

Known uses

Here are some known uses:

- Companies such as CrowdStrike and CyberArk provide a centralized management console from which administrators can connect to their enterprise network to monitor, protect, investigate and respond to incidents. This is accomplished by leveraging either an on-premises, hybrid, or cloud approach.

- Most of the large **cloud service providers** (**CSPs**) take care of the patching of physical servers from their side. Multiple patch management tools are in use, provided by different vendors such as Atera, NinjaOne, SolarWinds, SecPod, and ManageEngine.

Trusted compute patterns

Let's get started!

Problem

How to ensure that the software runs on a trusted system.

Context

For the cloud service consumer, there is a need to confirm their software is running on trusted hardware. From the CSP perspective, there is a need to confirm that the software allowed on their hardware is authentic. Hardware should have the capability to prevent any unsigned software from being run on the compute infrastructure. Malware and rootkits try to bypass initial security checks and try to launch themselves even before the operating system is launched. In this model, we look at how the **basic input/output system** (**BIOS**) on a cloud-based server environment can be protected from malicious code.

Solution

The **Trusted Computing Group** (**TCG**) has developed and promoted a technology called **trusted computing** (**TC**). The key idea of TC is to provide hardware control over which software can be permitted to run on it. The mechanisms required to support this pattern include **trusted platform modules** (**TPMs**), **hardware security modules** (**HSMs**), and **digital signatures** (**DSs**). Trusted compute platforms are made available by the cloud provider with TPMs, which are HSMs that enable security assurance by validating DSs of code, starting at the BIOS using a measured boot.

You can see a visual representation of this in the following diagram:

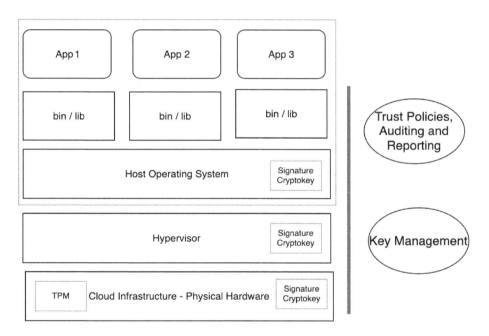

Figure 5.4 – TC pattern

In this model, consistency is enforced by leveraging a combination of hardware and software controls. The hardware loads a unique encryption key from the HSM that is inaccessible to the rest of the system, including the owner or provider of the infrastructure. This security validation from the silicon chip and upward, combined with remote monitoring of this platform, provides cloud consumers with a means to verify that their workloads are running on trusted compute platforms that meet their security assurance requirements.

A related topic is confidential computing, which will be discussed in the context of *data-in-use* protection. This pattern will be discussed in *Chapter 9*.

Known uses

Here are some known uses:

- IBM Cloud provides enhanced security verification of compute environments by using **Intel® Trusted Execution Technology (Intel TXT)**. TPM also known as **International Standards Organization/International Electrotechnical Commission (ISO/IEC)** *11889*) is an international standard for a secure cryptoprocessor (`https://en.wikipedia.org/wiki/Secure_cryptoprocessor`), a dedicated microcontroller designed to secure hardware using integrated cryptographic keys. TCG has certified TPM chips manufactured by Infineon Technologies, Nuvoton, and STMicroelectronics, having assigned TPM vendor **identifiers (IDs)** to Advanced Micro Devices, Atmel, Broadcom, IBM, Infineon, Intel, Lenovo, National Semiconductor, Nationz Technologies, Nuvoton, Qualcomm, Rockchip, Standard Microsystems Corporation, STMicroelectronics, Samsung, Sinosun, Texas Instruments, and Winbond (`https://en.wikipedia.org/wiki/Trusted_Platform_Module`).

- A reference implementation of a trusted compliant cloud based on technologies such as Intel TXT, TPM, virtualization by VMware, and cloud and data control by HyTrust is described in this document: `https://builders.intel.com/docs/cloudbuilders/Trusted-compliant-cloud-a-combined-solution-for-trust-in-the-cloud-provided-by-Intel-HyTrust-and-VMware.pdf`.

- Google provides shielded VMs so that when the VM boots, it's running code that hasn't been compromised. Shielded VMs provide a trusted firmware based on **Unified Extended Firmware Interface (UEFI)** and a **virtual trusted platform module (vTPM)** that validates guest VM pre-boot and boot integrity, and generates and protects encryption keys. Shielded VMs also include Secure Boot and Measured Boot to help protect VMs against boot- and kernel-level malware and rootkits. Integrity measurements collected as part of Measured Boot are used to identify any mismatches between the "healthy" baseline of the VM and the current runtime state.

Securing hypervisors

Let's get started!

Problem

The next key area to be discussed in the stack is how to protect and secure hypervisors.

Context

Virtualization and automation are the key enablers for the cloud. The virtualization layer sits between the physical infrastructure and the VMs. A hypervisor or *VM manager* is used to run numerous guest VMs and applications simultaneously on a single host machine and to provide separation between the guest VMs. As shown in the following diagram, virtualization provides a way to slice and dice physical infrastructure and provide this as VMs with variable configurations to the end users. The virtualization technology is delivered by hypervisors:

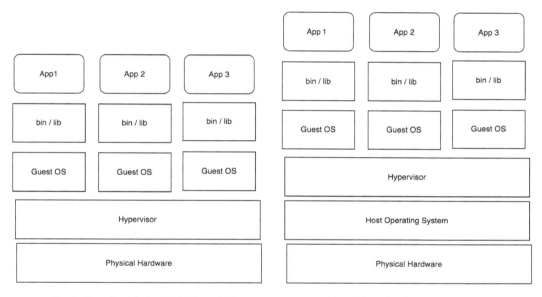

Figure 5.5 – Hypervisor types

As shown in the preceding diagram, there are two types of hypervisors. The *Type 1* hypervisor is a component that sits directly on top of the bare-metal infrastructure. In the case of *Type 2* hypervisors, there will be a host operating system layer in between the bare-metal and hypervisor. Depending on the type and the consumption model, the security management responsibility and protection of the stack vary. You can understand more about the hypervisor types by going through the links provided in the *References* section at the end of the chapter.

It is important to protect the hypervisor layer. A compromised hypervisor can affect all guest VMs running on top of it. An exposure or breach due to any vulnerability in this layer can expose data private to guest VMs. A hypervisor-based attack is an exploit in which an intruder takes advantage of vulnerabilities in the program used to allow multiple guest operating systems to share a common hardware processor. Here are some of the vulnerability and security threats applicable at the hypervisor layer:

- **VM escaping**–Hypervisors are designed to support strong isolation between the host and the VMs, but if there are vulnerabilities in the operating system running inside the VM, attackers can leverage that to break this isolation. In this situation where the isolation between VMs is broken, a guest VM might be able to control another VM or start communicating with the host operating system bypassing the hypervisor. Such an exploit opens the door to attackers to gain access to the host machine and launch further attacks. The VM can also control the compromised hypervisor and hackers can even move VMs.

- **Hypervisor management vulnerability**–A hypervisor is like an operating system or platform hosting multiple services including VMs on it. Users leverage the management consoles to control their VMs remotely. Any vulnerability in the client console or server side of the hypervisor management program can provide backdoor entry to a host or platform. This will provide easy access for hackers to break into the platform and control every VM. This type of security breach is of high risk to the virtual environment. If the management is done through any administrative machine, the hardening of that box is also important to prevent any illegal access. Any major failure of the hypervisor can result in the collapse of the overall system, including bringing down the guest VMs. Enforcing a security mechanism in a VM is an intermediate task. A virtualized environment is known for concurrently running multiple instances, applications, and even multiple operating systems, which again can be administered by multiple users in both a single-host and a multiple-host system with different user access. This environment also enables users to create, migrate, copy, and roll back VMs and run a multitude of applications. One of the other challenges is controlling a home or a guest operating system.

- **Sniffing and spoofing**–Sniffing and spoofing attacks work at the network interface layer. VMs interact through virtual switches to form a network. Sniffing is a passive attack to get some data. In this type of attack, VMs, if not restricted, can sniff out packets traveling to and fro to other VMs in the network or the whole network. Spoofing is an active security attack, sending improper requests for establishing a connection or intrusion from an untrusted source.

- **Hyperjacking**–Hyperjacking describes an attack in which a hacker takes malicious control over a hypervisor that creates a virtual environment within a VM host. The point of the attack is to target the host operating system by running applications on the VMs and the VMs themselves being completely oblivious to their presence.

- **VM theft**–An attacker can exploit security flaws to access information on a physical machine. With this access or information, the attacker can copy and/or move a VM in an unauthorized manner. This unauthorized copy or copy or movement of VM files can cause a very high degree of loss to a cloud consumer if the files and data contained in the VM are sensitive.

- **VM sprawl**–This is a situation of having many VMs in the virtual environment without proper control or management. In this scenario, a rightful VM may not get the required or requested system resources (that is, memory, disks, network channels, and so on). During this period, these resources are assigned or locked by other VMs, and they are effectively lost.

This is a scenario where the enterprise is leveraging bare-metal servers to set up its own private cloud and has control over the virtualization layer. Typically in most cloud scenarios, the cloud operator takes care of the hypervisor-layer hardening.

Solution

The same principles for securing bare-metal servers apply to hypervisor protection as well. The concepts of access control, hardening services at the hypervisor level, and vulnerability and patch management apply to the hypervisor level for the respective components. With regard to access control, it's advisable to restrict copy and move operations for VMs containing critical/sensitive information. Additionally, for this security solution pattern to prevent attacks on VMs and limit VM theft, the cloud provider will include a firewall and **intrusion detection system (IDS)** to secure and isolate VMs that are hosted on the same physical machine. The solution components are detailed in the following diagram:

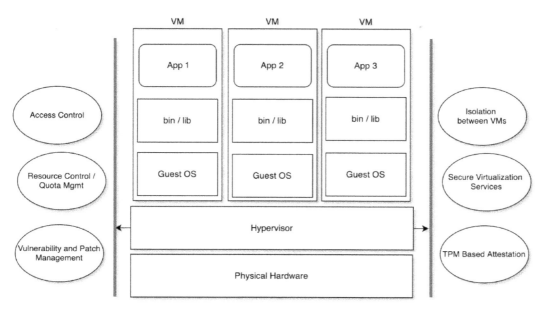

Figure 5.6 – Securing a hypervisor

Known uses

The vSphere **Security Configuration Guide** (**SCG**) is the baseline for security hardening of VMware vSphere itself, and the core of VMware security best practices. Started more than a decade ago as the VMware vSphere Security Hardening Guide, it has long served as guidance for vSphere administrators looking to protect their infrastructure.

Protecting VMs

Let's get started!

Problem

Understanding how to secure VMs.

Context

With virtualization, compute resources are made available in the cloud in the form of VMs. VMs are like a server environment created within a computer. They have a guest operating system. The management plane of the hypervisor enables us to create and run multiple VMs. All the threats that are relevant for bare-metal servers are also applicable to VMs.

You can see a visual representation of this in the following diagram:

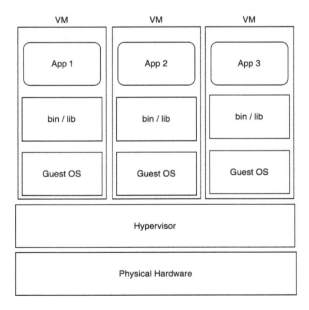

Figure 5.7 – VMs

VMs are subjected to the following attacks, in addition to those that are applicable to bare-metal servers as well:

- A VM can get infected with malware or operating system rootkits at runtime. This is malicious software that gives unauthorized access to a computer. It is hard to detect and can conceal its presence within an infected system. Hackers use rootkit malware to remotely access the VM and manipulate and steal data from it.

- A malicious VM can potentially access other VMs through shared memory, network connections, and other shared resources. Attacks from the host operating system and/or co-located VMs are known as outside-VM attacks. These attacks are difficult to detect. In the virtual environment, multiple VMs are usually provisioned on the same physical server in a cloud environment. VMs sharing the server raise potential threats through cross- or covert channel attacks. In this side-channel attack, a malicious VM breaks the isolation boundaries between VMs to access the shared hardware and cache locations to extract confidential information related to the target VM.

- VM image sprawl is a situation caused by the massive duplication of VM image files. In this, a large number of VMs created by cloud consumers may cause enormous administration issues if the VMs are infected and need security patching.

Solution

The hardening of a virtual server is similar to bare-metal server hardening. Hardening is the process by which the security of the virtual server is improved by reducing its exposure to threats and vulnerabilities.

You can see a visual representation of this in the following diagram:

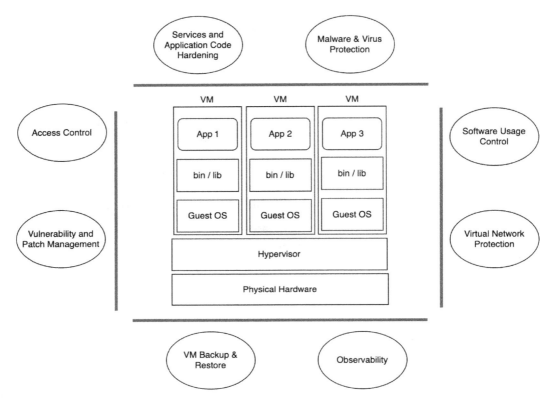

Figure 5.8 – Securing VMs

Hardening includes the following:

- Removing any unnecessary usernames and logins.

- Removing unnecessary software.

- Removing any unnecessary drivers.

- Disabling any unwanted or unused services.

- Patching the VMs to bring the software up to the latest levels without any security vulnerabilities.

- Disabling/removing all unnecessary virtual hardware (**central processing units (CPUs)**, **random-access memory (RAM)**, and media devices) and limiting any direct access or VMs utilizing physical resources.

- Performing integrity validation, signature checking, or virtual encryption (on the guest operating system/server) to prevent unauthorized copying.

- Observing and monitoring change management procedures, remote auditing/control, and important statistics related to VM health. This includes the CPU utilization and overall bandwidth of the VMs, as well as the physical server, to ensure good health. In cases where the underlying infrastructure is not able to provide the capabilities needed to fulfill the VM demand, automation must be in place to migrate VMs across hosts to support the application demand.

- Depending on the shared security responsibility model in the cloud, the cloud consumer may be responsible for maintaining the physical infrastructure, its updates, and security patches.

Known uses

Here are some known uses:

- The tools used for VM hardening are similar to those listed in the known uses under bare-metal protection.

- Microsoft Defender for Endpoint provides industry-leading endpoint security for Windows, macOS, Linux, Android, iOS, and network devices. It's delivered at a cloud scale, with built-in **artificial intelligence** (**AI**). It enables the discovery of all endpoints and offers vulnerability management, endpoint protection, **endpoint detection and response** (**EDR**), mobile threat defense, and managed hunting, all in a single, unified platform.

- Symantec provides endpoint protection that goes beyond antivirus, blocking advanced threats as well. Traditional antivirus does not play well in virtualized infrastructure, so some vendors such as McAfee provide antivirus optimized for virtual environments. This optimizes anti-malware protection for virtualized deployments, freeing hypervisor resources while ensuring up-to-date security scans are run according to policy.

Securing containers

Let's get started!

Problem

Patterns for securing containers.

Context

Containers provide a better way to efficiently use the underlying infrastructure compared to VMs. Application components and all dependencies are packed inside a container and executed in a secure way.

As shown in the following diagram, containers do not have any guest operating system. Instead, the container leverages the operating system and environment of the underlying layer:

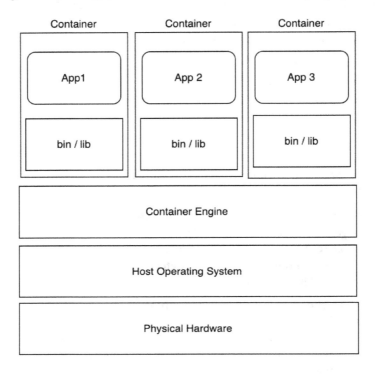

Figure 5.9 – Containers

Containers bring several advantages, important one being *build once, run anywhere*. This is achieved by packing everything that an application needs into a container, thus isolating the application from the server on which it is running. A containerized application has everything it needs, packed as a container image. A container runtime (also known as **container engine**, which is a software component deployed on a host operating system) is needed to run containers. This image can be run on any machine, such as on a laptop or on a server in a cloud environment that has the container runtime deployed. Containerized applications can be deployed across a cluster of servers, leveraging container management platforms such as Kubernetes to automate this process. The security threats in a containerized environment are similar to deployments in a traditional environment. However, there are several changes in the way applications are run as containers. If we take a deeper look at the container threat model, there are several internal and external attackers involved, such as the following:

- External attackers include people or processes trying to gain access to deployments or images from outside

- Internal attackers are malicious insiders such as developers or administrators who have privileged access to the deployment as well as inadvertent actors who may have caused problems because of incorrect configuration

The routes to attacking a container-based deployment include the following:

- **Exploiting vulnerable application code**—The application typically uses several insecure or outdated libraries and packages. These tend to have many vulnerabilities that an attacker can exploit. Similar to any other application development project, badly written code is a security risk with container-based deployment as well, which can expose the enterprise to attack.

- **Gaining privileged access**—Containers provide an easy way to package the application and its dependencies, but many times, the creator of the container image may not pay sufficient attention to configuring the container image correctly. This can introduce weaknesses that an attacker can leverage to gain privileged access to the container instance. Container development includes an important step of build activity. Attackers may try to inject a malicious library or code injected in the build phase to leverage as a backdoor to gain access to the container in production.

- **Insecure container registry**—Builds or container images are shared for deployment leveraging container registries. Securing the container registry and controlling access to the images as part of the **continuous integration/continuous deployment** (**CI/CD**) process is an area of concern.

- **Poor configuration and secrets exposure**—Container deployment requires setting up secrets, ports, and underlying networks correctly. Credentials are shared with container code through several mechanisms in a containerized deployment. This is an area prone to attack by hackers. Also, if the network configuration is not set up correctly, this is another key area of vulnerability. Some examples of these secrets include passwords used to access databases, user-generated passwords, or **application programming interface** (**API**) keys/credentials as well as **Secure Shell** (**SSH**) keys or certificates.

- **Container runtime vulnerabilities**—As discussed in the introduction of this section, containers are run on top of a popular container runtime or container engine, such as `runC`, `docker`, `containerd`, or CRI-O. Many of these container runtimes meet the specifications outlined by the **Open Container Initiative** (**OCI**). Following this standard means the runtimes are matured and hardened to provide robustness and portability. OCI compliance also means interoperability for container registries. Containers are not first-class objects of the operating system and leverage *Linux kernel primitives: namespaces* (who you are allowed to talk to), **control groups** (**cgroups**: the number of resources you are allowed to use), and **Linux Security Modules** (**LSMs**)—what you are allowed to do. These kernel primitives allow you to set up secure, isolated, and metered execution environments for each process running inside the container. There is a possibility that there can be bugs in these primitives that make the container runtime vulnerable. Such vulnerabilities can be fatal if they allow the container to escape the strong isolation mechanisms to access the host operating system and other containers.

- **Container management**—Container deployment, orchestration, and management are done with tools such as Kubernetes. Complex deployment leveraging these tools involves multiple configurations and permission management. If this is not set and managed correctly, it becomes a target for attackers.

Solution

A container security pattern addresses the challenges discussed in the preceding context through following these best practices:

- **Hardening the container environment**—The container environment consists of the host operating environment, the container runtimes, and the related services including the container registry. The administrator needs to ensure that only the required services are up and that least-privilege access is provided to the users/systems based on the need. Avoid or shut down any other additional services running on these servers that are not required for the function, and limit container capabilities to what's absolutely necessary to reduce the attack surface. The administrator also needs to lock down the container privileges to reduce and prevent any escalation attacks. Configuration and security reviews for both the containers and the underlying systems need to be done regularly.

- **Keeping the container platform updated**—Keep the container platform updated with patches to ensure the container runtimes are free from vulnerabilities. Updating and strengthening all runtimes and container management tools against attack should be performed regularly.

You can see a visual representation of such security processes in the following diagram:

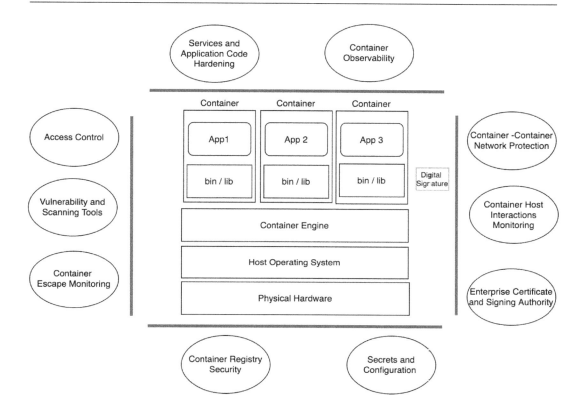

Figure 5.10 – Container security

- **Ensuring the enterprise only uses trusted containers**—The next step is to ensure that the enterprise is only using trusted containers and downloads container images from trustworthy repositories. The recommendation is to deploy from private registries rather than directly deploying container images from untrusted sources. The organization needs to validate that the container image is free of vulnerabilities and does not have any outdated or insecure packages or components. It is important to put into practice a process to review reused internal components to ensure that they are safe and not compromised so that container images are secure. Roll back or replace high-risk containers with patched and updated ones. Bad containers can be thrown away and replaced with new ones without vulnerabilities, and this is different from the VM hardening pattern discussed earlier. Containers are immutable objects, so patch updates take the form of replacing the previous version with an updated one. It is recommended to reduce the container size and life where possible. Unused containers should be shut down and removed for improved security. Short-lived containers are a secure design practice that can be adopted by enterprises.

- **Automating vulnerability scanning**—It's critical to have visibility into a container's entire life cycle, from build to deployment. Automated vulnerability scanning is recommended to ensure no insecure container makes it into production. The scanning process to detect any vulnerability in containers cannot be a one-time activity, so this needs to be integrated with the CI/CD pipeline. This way, whenever the content is changed or updated, security scanning is performed. Also, when running containers go out of date with regard to packages, they can be updated for security patches.

- **Network segmentation**—Network segmentation is important, and isolating the containers as per the business requirement is a good practice to reduce the blast radius in case of an attack. This will limit the attack or compromise through the network layer to the minimum number of containers. Restrict container-to-container communication to the required minimum and protect the required interactions with a clear segmentation of the network and firewall rules. Also, auto-scaling can result in more containers being spun up, at least at times, so care needs to be taken so that security doesn't break that. Container network traffic needs to continuously be monitored for any abnormal behavior and suspicious activities. The network configuration needs to be penetration-tested to ensure security is enforced at all layers. Network security patterns will be discussed in detail in the next chapter.

- **Ensuring security**—The shift toward containers has given developers more responsibility for security. Consequently, developers will need to spend time learning what trusted container image development really means, how to properly vet code for security, and how to adopt continuous security assessment tactics and minimize the attack surface of the overall container by removing unnecessary components. To truly keep containers secure, developers also need to have **end-to-end** (E2E) visibility into their entire container environment, allowing them to immediately detect and block malware and other security threats before they cause significant damage. With strong attention to some basic security best practices and benchmarks, administrators can make containers as secure as any type of code. Today, most developers focus on the functionality and stability of their code. They also need to take guidance from their security focal to ensure their code runs on secure containers as well.

- **Regular auditing**—Regular auditing of the container environment, leveraging security benchmarking and hunting tools, is recommended.

Known uses

Here are some known uses:

- Palo Alto Networks provides a solution for container security that covers scanning images to identify high-risk issues, prevent vulnerabilities, provide developers with trusted images, and gain runtime visibility into various containerized environments.

- Aqua provides a solution covering different aspects of container security that can be integrated with the CI/CD pipeline. This includes CI scans, Dynamic Analysis, Image Assurance,

Risk Explorer, Vulnerability Shield, Runtime Policies, behavioral profiles, workloads firewall, secrets injection, auditing, and forensics.

- Red Hat Advanced Cluster Security for Kubernetes integrates with **development operations** (**DevOps**) and security tools to help mitigate threats and enforce security policies that minimize operational risk to containerized applications.

Securing serverless implementations

Let's get started!

Problem

Determining what the patterns for securing serverless deployments or cloud functions should be.

Context

Serverless, as shown in the following diagram, is the latest among the computing type options. It is also called **Function as a Service** (**FaaS**). **Amazon Web Services** (**AWS**) Lambda, Azure Functions, IBM, and Google Cloud Functions are popular examples of serverless computing models. In this model, an application is broken into separate functions that run when triggered by some action. The consumer is charged only for the processing time used by each function as it executes:

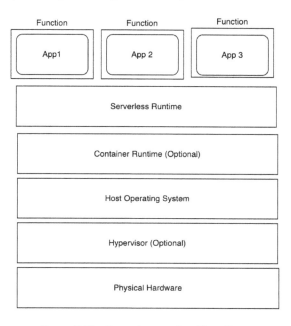

Figure 5.11 – Serverless or cloud functions

A major challenge with securing serverless functions is that they are short-lived or ephemeral. So, it can be challenging to monitor and detect malicious activity in serverless functions. Serverless functions rely on underlying components and other cloud services to execute the tasks. Cloud providers only provide limited visibility compared to containers on the underlying infrastructure, so it can be difficult to identify where and when a function runs. For example, two invocations of the exact same function can run on completely different nodes. These dependencies can become complex and are hard to identify and harden for security. It is critical to ensure cloud functions are secure as they are heavily reused across applications. If there is any vulnerability with the function, it can impact multiple services and applications. Any prominent security issues should be remediated automatically. Vulnerabilities can exist in the underlying server and runtime infrastructure as well as in the code, dependencies, and third-party libraries used by the function. An attacker can exploit any of these vulnerabilities to gain access to cloud resources.

Another possible area to lock down is permissions and configuration settings for the functions. Most often, the functions are overprovisioned in terms of access in dev/test environments. More permissions are granted than what's required to perform the task. Overprovisioned function permissions are high-risk, as if a hacker gains access to the function, they can harm other services.

Solution

As serverless is a new model for executing code, securing serverless also requires a paradigm shift. Traditional patterns of **defense in depth** (DiD) with firewalls and surrounding applications may not be effective for securing serverless. The organization must additionally build security around the functions hosted on cloud provider infrastructure. Hardening and securing the underlying infrastructure used by the cloud function is performed by the cloud provider. It's worth noting that in a true hybrid model, serverless computing can end up being implemented extending across one or more clouds as a set of microservices that cause resources to be started and, hopefully, terminated upon need.

You can see a visual representation of how to secure serverless and cloud functions in the following diagram:

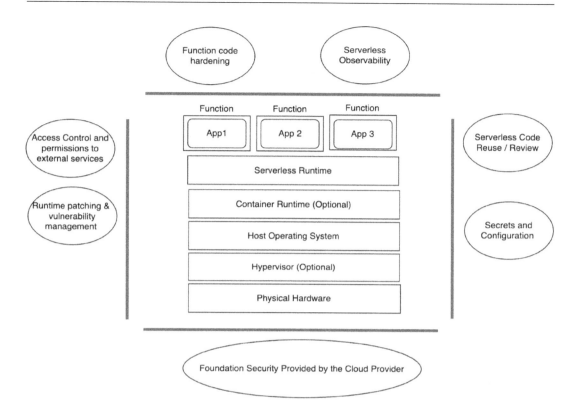

Figure 5.12 – Securing serverless or cloud functions

The key elements and tasks in this pattern for securing serverless functions include the following:

- Ensuring proper hardening of the function to enforce the least-privilege access control principle so that each function does no more and no less than what is required to execute the task.

- Scan the application code executed as part of the cloud function for any code, package, or library vulnerabilities and malware. Also, scan dependent libraries that are indirectly used by the serverless function.

- Ensure that the serverless environment is free from known security risks through continuous discovery, scans, and monitoring.

- It is important to check and validate the configuration to ensure no sensitive data, secrets, keys, or confidential elements are exposed through cloud functions.

Known uses

- AWS discusses a shared responsibility model for serverless security in which customers are free to focus on the security of application code, the storage and accessibility of sensitive data, observing the behavior of their applications through monitoring and logging, and **identity and access management (IAM)** to the respective service. AWS will take care of server management and deliver flexible scaling with automated **high availability (HA)**.

- Aqua provides tools for the secure execution of serverless functions. The tools ensure least-privilege permissions, scan for vulnerabilities, automatically deploy runtime protection, and detect behavioral anomalies.

- Imperva Serverless Protection embeds itself into the function in order to defend against new attack vectors emerging in serverless functions.

- CloudGuard's solution uses a code-centric platform that automates security and visibility for cloud-native serverless applications from development to runtime. By analyzing the serverless application code before and after deployment, organizations can achieve a continuous serverless security posture—automating application hardening, minimizing the attack surface, and simplifying governance.

Summary

In this chapter, we learned about the various patterns to secure the different types of compute infrastructure. We provided the context to the types of attacks that can happen on different layers of the compute infrastructure. For different compute types of the modern hybrid cloud infrastructure—namely, bare-metal servers, VMs, containers, and serverless—we discussed the patterns that can be used to address their security.

In the next chapter, we will learn about the patterns to secure a hybrid cloud network, which is an essential component of hybrid cloud infrastructure.

References

Refer to the following resources for more details on the topics covered in this chapter:

- IBM bare-metal servers using Intel® Trusted Execution Technology (Intel TXT)— `https://cloud.ibm.com/docs/bare-metal?topic=bare-metal-bm-hardware-monitroing-security-controls`

- Virtual Trusted Platform Module for Shielded VMs— `https://cloud.google.com/blog/products/identity-security/virtual-trusted-platform-module-for-shielded-vms-security-in-plaintext`

- CyberArk endpoint protection, privilege management, and patching— `https://www.cyberark.com/products/endpoint-privilege-manager`

- Microsoft Defender for Endpoint—`https://docs.microsoft.com/en-us/microsoft-365/security/defender-endpoint/microsoft-defender-endpoint`

- McAfee Management for Optimized Virtual Environments AntiVirus—`https://www.mcafee.com/enterprise/en-us/assets/data-sheets/ds-move-anti-virus.pdf`

- Palo Alto Networks container security— `https://www.paloaltonetworks.com/prisma/cloud/container-security`

- Aqua container security—`https://www.aquasec.com/products/container-security/`

- Red Hat Advanced Cluster Security for Kubernetes— `https://www.redhat.com/en/topics/security/container-security`

- Aqua Security for Serverless Functions (FaaS)— `https://www.aquasec.com/products/serverless-container-functions/`

- Imperva serverless security—`https://www.imperva.com/products/serverless-security-protection/`

- AWS shared security model for serverless— `https://aws.amazon.com/blogs/architecture/architecting-secure-serverless-applications/`

- Check Point CloudGuard serverless self-protection— `https://www.checkpoint.com/cloudguard/serverless-security/`

6

Implementing Network Isolation, Secure Connectivity, and Protection

In this chapter, we will learn about the patterns that can be leveraged to secure the hybrid cloud network infrastructure. Along with compute and storage, the network is one of the core infrastructure components. To scale deployments across multiple clouds, the network needs to be scaled and managed as a core unit. This is achieved through network virtualization, which makes sharing physical network resources possible by creating a virtual network. This is enabled by combining networking resources across internal and external networks. In this chapter, we will discuss the foundation patterns for managing this large virtual network. The patterns fall under the following categories:

- Network isolation – how to isolate network traffic based on purpose and target environments

- Network connectivity – how to establish secure connectivity in hybrid and multi-cloud environments

- Network protection – how to implement network protection for cloud workloads

The following diagram illustrates the patterns:

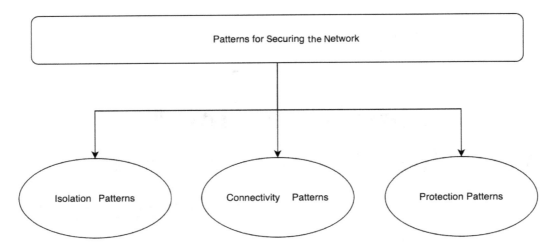

Figure 6.1 – Patterns for securing cloud network infrastructure

In this chapter, we will cover the following main topics:

- Network isolation patterns
- Secure network connectivity
- Network protection

Network isolation patterns

Let's get started!

Problem

Why is network isolation needed and how can we achieve network isolation?

Context

Most organizations' applications are distributed across different data centers and multiple cloud environments based on business needs. These applications need to connect to other cloud and on-premises applications to fulfill their functions. Networks are currently protected with perimeter-based network security. This mostly enables what is called **north-south** (client to server) traffic and is typically done by leveraging network firewalls. But applications or assets within the network are generally trusted, and **east-west** traffic is allowed without any controls. This traditional approach of network protection cannot be extended to the cloud because if the network gets attacked, it exposes the entire internal and external network that is spread across the cloud. Modern security thinking, in fact, says that you should assume that your network is compromised, meaning that the chances are very high that an attacker has already been able to penetrate your network perimeter defenses.

To protect your network and limit the range of damage that a bad actor can do when access to a network is granted, you must take precautions to ensure workloads are as isolated as possible to limit the number of apps and worker nodes that are publicly exposed. To be effective in cloud environments, we need a mechanism to limit the blast radius for any bad events. Traditional segmentation approaches are inadequate as they designed and deployed networks aligned to business and application workloads. But in a cloud-native architecture, resources and microservices are shared across different applications. Also, in cloud shared responsibility models, the cloud provider handles most of the physical hardware network resources (routers, switches, and so on) while the cloud consumer needs are different for establishing network isolation.

Solution

The public network handles public traffic to hosted applications or online resources. A server or application with public network connectivity means the hosted applications or online resources are provisioned and attached to a public IP address and are reachable through the internet.

A private network provides free, secure connectivity between resources hosted on the cloud. The network is typically provided by the cloud provider and is unmetered and free within a region or zone so that customers can easily move as well as connect to their resources securely in a private network.

A management network is one that is separated from workload- or application-specific inter-server traffic and primarily used for out-of-band management.

Traditionally, cloud providers used a single flat network that was shared across customers. A cloud provider gives basic, customer-level network isolation through individual **virtual local area networks (VLANs)**. Cloud providers provide customers with the ability to further isolate their networks or systems from one another based on their needs through network segmentation.

Network segmentation describes the approach to dividing a network into multiple subnetworks. This involves grouping related applications and data into different sets and running them in a specific subnetwork. Applications in one subnetwork cannot see or access apps in another subnetwork. Network segmentation limits the access that is provided to internal and external applications. For instance, this isolation of network traffic can be based on the application function or on the environment such as development, test, or production. Network segmentation can be also performed based on data sensitivity and regulatory requirements such as PCI and HIPAA.

Cloud providers leverage micro-segmentation to manage network access between the subnetworks. Micro-segmentation is a method for managing network security in which administrators implement security policies to limit traffic based on the principles of least privilege and limiting access to select network segments. This helps not only to reduce the attack surface but also, in the event of an incident, to contain the breach to a limited radius as well as improve and strengthen regulatory compliance.

The cloud consumer can achieve network segmentation through leveraging virtual gateways and routers to further segment the VLAN.

Typically, the cloud provider assigns or associates two VLANs (a public VLAN and a private VLAN) to every public-facing server. VLAN spanning needs to be implemented when the customer has servers or workloads across multiple data centers and they need to talk to each other. While VLANs are an OSI Layer 2 construct that can provide isolation for given customers' traffic, further isolation is often needed to implement *zones* or *tiers* to restrict workload-specific traffic in the private network. OSI Layer 3 isolation (based on subnets and IP addresses) is achieved using gateway routers.

Most cloud providers now provide what's called a **Virtual Private Cloud** (**VPC**) that allows running instances logically isolated to their cloud account. With a VPC, cloud resources are provisioned into a virtual network that the client defines, and it resembles a dedicated traditional network like your own data center. In this model, the customer has the ability to select their own IP address space, public and private subnet configuration, and management of route tables and network gateways.

To understand the various levels of network isolation that can be achieved within a VPC, we need to understand a few of the constructs detailed in the figure that follows:

- Region
- **Availability Zone (AZ)**
- Internet gateway
- NAT gateway
- Network ACL
- Security groups

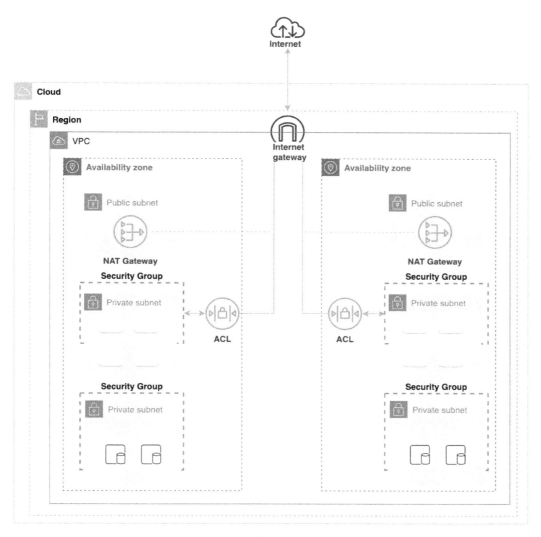

Figure 6.2 – Network isolation

In the cloud, a **region** refers to a separate geographic area with a completely independent physical infrastructure. Isolation within each region is achieved through locations known as AZs. Each region may have multiple AZs. Each AZ has one or more physical data centers that are fault-isolated from other AZs in the region. The AZs within a region are connected through low-latency networking links. To achieve high availability for an application, the application is deployed across multiple AZs. VPCs can span multiple AZs.

The resources inside a VPC can communicate with the internet through an **internet gateway** component. If the resources, such as the compute instance, have a public IP address in the public subnet, the gateway enables them to connect to the internet. Also, if an external resource (such as a remote client) wants to initiate a connection through the internet to the cloud resources, they come through an internet gateway. The internet gateway looks up the VPC route tables for internet-routable traffic and sends the traffic to the right target leveraging a NAT gateway component.

A **NAT gateway** is a **Network Address Translation** (**NAT**) service. This is needed so that resources or instances in a private subnet can connect to services outside the VPC. For outbound connectivity for instances in private subnets, we need to create a NAT gateway in a public subnet. The NAT gateway is created with a public static IP address at the time of creation. Similarly, these resources in one VPC can connect to other VPCs or on-premises resources through a private NAT gateway. A private NAT gateway routes traffic through a transit gateway or a virtual private gateway.

The NAT gateway replaces the source IP address of the instances with the IP address of the NAT gateway. When sending response traffic to the resources, the gateway device translates the addresses back to the original source IP address.

For high availability, NAT gateways are created in multiple AZs and configured to route traffic to the resources. There are limits on the volume and throughput of traffic that a NAT gateway can handle. For scalability and preventing packet loss, network-heavy resources need to be split across multiple subnets and NAT gateways need to be created for each subnet.

By default, all inbound and outbound network traffic is allowed into the subnet inside the VPC. A network **Access Control List** (**ACL**) is an additional or optional layer of security within a VPC that acts as a firewall for controlling traffic in and out of one or more subnets. A network ACL can be created and associated with a subnet to control the inbound and outbound traffic. A network ACL can specifically separate inbound and outbound rules to either allow or deny traffic. By default, a custom network ACL denies all inbound and outbound traffic. If a subnet is not associated with a network ACL, the subnet is automatically associated with the default network ACL. A network ACL is made up of a numbered list of rules. The rules are evaluated in ascending order to determine whether traffic can be allowed in or out of any subnet. Network ACLs are stateless, which means that responses to allowed inbound traffic are subject to the rules for outbound traffic (and vice versa).

A **security group** is like a firewall that can control the network traffic to any cloud resource. Instances such as compute or web can be associated with a security group. The inbound and outbound traffic for the instances is determined by rules specified in the security group. Like the network ACL, there are separate sets of rules for inbound traffic and outbound traffic. But unlike the network ACL, security groups are stateful. So, if the rules allow sending a request from an instance, the response for that request is allowed to reach the same instance regardless of the inbound security group rules. Similarly, responses to inbound traffic are allowed to leave the instance, regardless of the rules specified. When we create a VPC, a default security group is created.

By leveraging all these components, network isolation and micro-segmentation can be achieved at various levels. Customers implement network segmentation based on various models. For instance, based on the type of environment, networks for production, staging, and development environments are different. Audit and regulatory requirements also mandate certain workloads to be running in isolated environments. A network can be separated for different business units and domains such as finance, HR, or sales, and applications are provisioned in those specific networks. Even within a single application, separation can be achieved between various tiers such as frontend or web, backend APIs, and data. Network isolation minimizes the security risk by reducing the attack surface. Even if hackers breach the one-layer defense, their access is limited to a network segment. By organizing the network into segments, we can also monitor the network better and handle the incidents and threats effectively. Segmentation can also help improve network performance by limiting traffic to certain zones as per the need and reduce congestion.

Known uses

The following are the known uses:

- Azure Virtual Network provides features to run **virtual machines** (**VMs**) and applications in an isolated and highly secure manner using private IP addresses. It also provides features to define subnets and policies to control access. An Azure Virtual Network NAT gateway can be leveraged to provide outbound internet connectivity for virtual networks (`https://azure.microsoft.com/en-in/services/virtual-network/`).

- An Amazon VPC provides a virtual network that users can leverage to create AWS resources in the subnets. Security and isolation can be achieved through security groups and a network ACL. Internet traffic and communication with resources in the public and private subnet are managed through internet and NAT gateways (`https://docs.aws.amazon.com/vpc/index.html`).

- A Google VPC provides networking functionality to provision Compute Engine VM instances, **Google Kubernetes Engine** (**GKE**) clusters, and the App Engine flexible environment. A VPC network is a global resource that consists of a list of regional virtual subnetworks (subnets) in data centers, all connected by a global wide area network. VPC networks are logically isolated from each other in Google Cloud (`https://cloud.google.com/vpc/docs/overview`).

- The VPC offering provided by IBM offers the ability to define and control a virtual network that is logically isolated from all other public cloud tenants. Logical isolation is implemented with virtual network functions and security features that give the customer granular control over which IP addresses or applications can access particular resources (`https://www.ibm.com/cloud/learn/vpc`).

Secure network connectivity

Let's get started!

Problem

How to ensure secure connectivity to multi-cloud and on-premises resources.

Context

In the current scenario, applications need to securely connect to resources spread across on-premises data centers and the cloud. The connectivity to the cloud may be established through a *front channel* such as through an internet service provider or through a back channel or direct link between the cloud provider and the customer's on-premises data center. Connectivity is also needed between VPCs for applications or business processes to be completed. The VPCs could be within the same account or different accounts of a customer. Certain use cases require connectivity between VPCs from the customer account to VPCs in the partner accounts as well. Some use cases require connectivity to the services or resources over a private network without having the need to go through the public internet. In a hybrid scenario, we also have requirements to connect from one cloud to another cloud.

Solution

Secure network connectivity is typically achieved in one of two ways: a direct connection to the cloud provider, or an internet-based routing to a cloud provider-hosted environment. A direct connection can be provided directly from the cloud and terminate at the customer's private network.

This model is described in the following figure. Sometimes there may be also a partner or ISP router in between, making the Direct Connect possible. The Direct Connect router at the cloud end connects to customers' public and private endpoints inside the VPC, as well as providing virtual interfaces to public cloud services. This means that there is a faster and more secure path to the data center than going through the public internet.

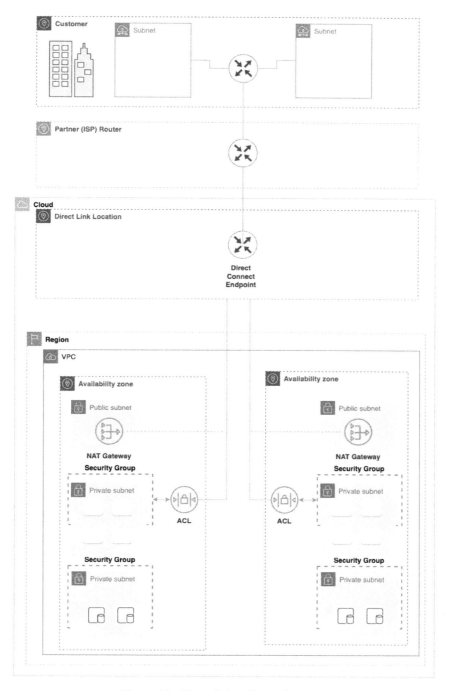

Figure 6.3 – Direct link or Direct Connect

In the case of customers who do not want to invest in costly Direct Connect solutions, a **Virtual Private Network** (**VPN**) is an option. A VPN securely connects the customer network to the cloud VPC network through an IPsec connection. Traffic between the two networks is encrypted by one VPN gateway and then decrypted by the other VPN gateway. This model, as described in the following figure, is called a client VPN and protects the data sent over the internet:

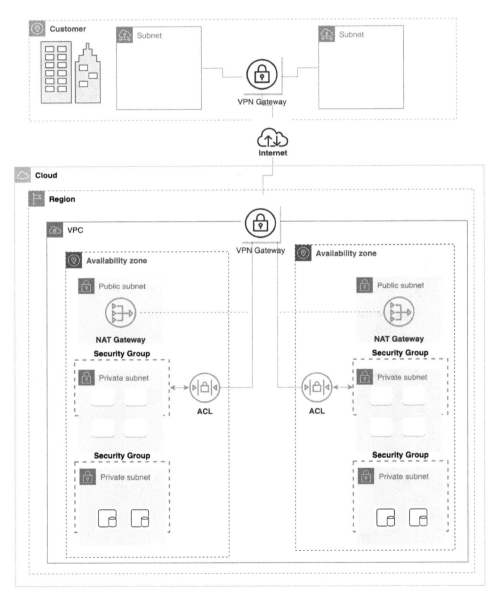

Figure 6.4 – VPN gateway

This is an easier and cheaper option to establish secure connectivity than the Direct Connect options. A VPN connection from a client location to the private network allows for unlimited file transfers, out-of-band management, and server rescue through an encrypted VPN tunnel. The VPN connection is established by using encryption such as SSL, PPTP, or IPsec. Access to the resources or instances is via its private 10.x.x.x IP address by using SSH or RDP. You can also connect to the management interface for maintenance and monitoring purposes. VPN connections can also be brokered through third-party internet-facing gateways that provide VPN termination and transfer traffic off the public network and move it to the private network. In this case, the VPN gateway (at least) must be internet-routable and reachable; no other servers need to be exposed to the internet. You can also use a VPN hub to connect multiple instances of VPN to communicate with each other.

Another alternative for establishing secure connectivity between VPCs is through VPC peering. As shown in the following figure, VPC peering allows you to make a connection to another VPC in the same cloud, same or different accounts, and across different regions. For an inter-region VPC peering connection where the VPCs are in different regions, the request must be made from the region of the requester VPC. To activate the connection, the owner of the accepter VPC must accept the VPC peering connection request.

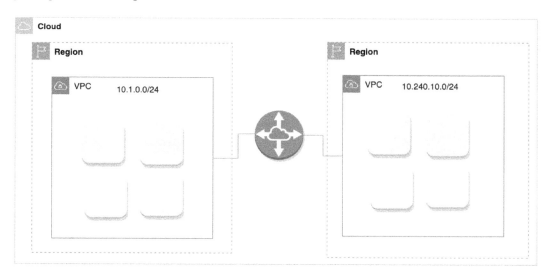

Figure 6.5 – VPC peering

Beyond a certain point, using several peer-to-peer connections becomes difficult to manage. That's when we need a Transit Gateway solution, as shown in the following figure:

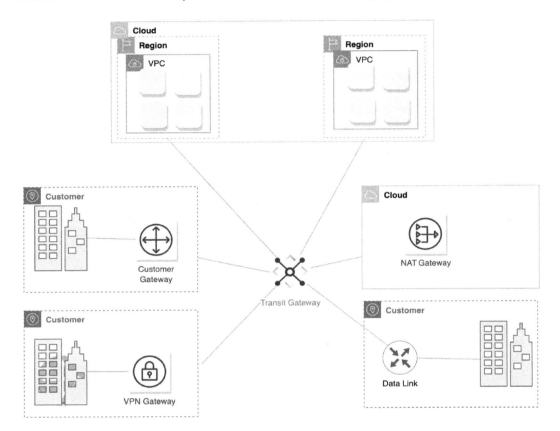

Figure 6.6 – Transit Gateway

Transit Gateway helps to connect several VPCs and also helps to easily manage traffic routing without having to maintain massive route tables for peer connections. Transit Gateway also seamlessly connects across regions, VPCs, and inline network security components, as well as to the customers' on-premises routers. Transit Gateway also provides a multicast feature that allows connecting to the network without having to make changes to the application or on-premises network. Transit Gateway can also route traffic to and from the VPC or VPN, thus providing a single control point for network monitoring and management.

Known uses

The following are the known uses:

- AWS Direct Connect provides connectivity over a dedicated network connection optic cable from a customer's location to the AWS cloud. An AWS Direct Connect location provides access to AWS in the region with which it is associated (`https://docs.aws.amazon.com/directconnect/latest/UserGuide/Welcome.html`).

- IBM Cloud Direct Link allows customers to connect their on-premises or collocated infrastructure directly to the IBM Cloud private network. Physical connections are available in IBM Cloud network points of presence around the world. No traffic across your Direct Link and between your servers touches the public network or otherwise interferes with your public network traffic (`https://www.ibm.com/cloud/direct-link`).

- Google's Dedicated Interconnect provides direct physical connections between customer on-premises networks and Google's network (`https://cloud.google.com/network-connectivity/docs/interconnect/concepts/dedicated-overview`).

- Azure ExpressRoute provides dedicated private network fiber connections to Azure (`https://azure.microsoft.com/en-in/services/expressroute/`).

- IBM Cloud provides the option for IBM Cloud-managed VPN endpoints to provide access to customer private networks over SSL, PPTP, or IPsec VPN gateways. This access takes the customer directly to their private network so that they can access and manage server infrastructure independently of the operating system (`https://www.ibm.com/cloud/vpn-access`).

- The IBM Cloud Network Gateway enables the custom routing designs needed to build a customer's own tiered or zoned network within IBM Cloud-provided VLANs (`https://www.ibm.com/cloud/network-appliances`).

Network protection

Let's get started!

Problem

How to provide network-level protection for resources.

Context

Protecting your resources at the network level is imperative for protecting cloud-based workloads. To provide comprehensive network protection, you might need to place purpose-built firewalls or multifunction appliances at appropriate locations based on the network design and deployment of your workload. The devices may vary based on the network layer that needs to be protected. At the cloud service or application layer, workloads are also subjected to **Distributed Denial-of-Service (DDoS)** attacks. There is a need to allow only good traffic to the multi-cloud resources and block bad traffic from reaching the servers.

Solution

Network firewalls help prevent malicious network packets from reaching the cloud resources. As shown in the following figure, cloud resources and services must define and implement the firewalls with appropriate policies to allow or deny a given workload. Network firewalls can run stateless and stateful traffic inspection rule engines.

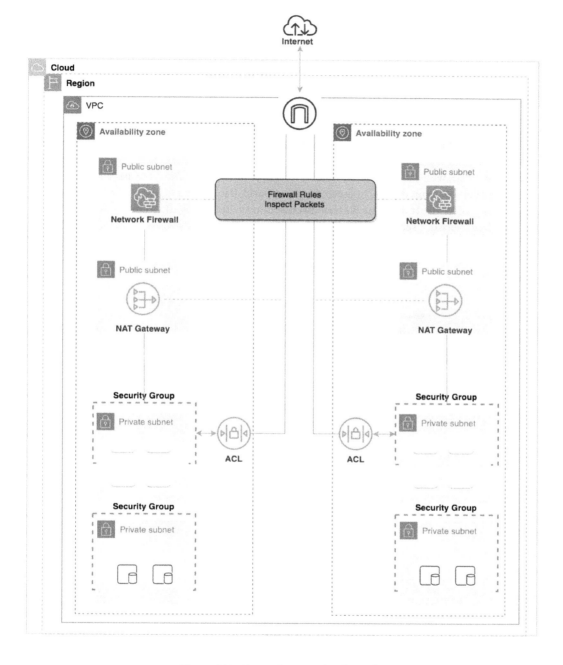

Figure 6.7 – Network protection firewalls

The engines use rules and other settings that are configured inside a firewall policy. Firewalls can be deployed on a per-AZ basis in a VPC. The firewall endpoint that filters the traffic can protect all of the subnets inside the zone except for the one where it's located. You can configure or define the rules to filter traffic appropriately. By blocking unwanted traffic into your environment, you can improve the performance of your entire system's network availability, workload, and security. Cloud providers' firewalls are available in the software and virtual appliance form as well as in the hardware form.

Traditionally, **intrusion detection systems (IDSs)** and **intrusion prevention systems (IPSs)** are protection systems at the network level. These systems analyze network traffic for specific signatures, behaviors, or anomalies to find any intrusion attempt. An IDS is designed to detect a potential intrusion and generate an alert, while an IPS goes beyond detection to block the intrusion by performing actions based on a set of preconfigured rules. In the multi-cloud world, it is difficult to insert a box in the middle to inspect and protect the traffic. The pattern discussed in the following diagram is leveraged to put the purpose-built network security devices on the edge or at the ingress/egress points of a workload:

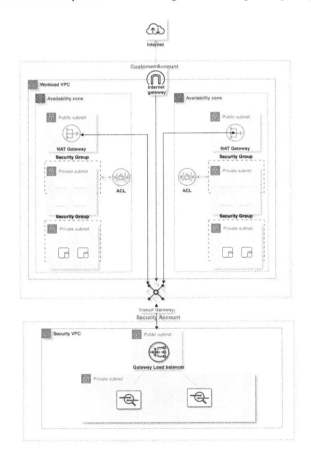

Figure 6.8 – Network protection – IPS/IDS

An **Application Load Balancer** (**ALB**) is another component that operates at the OSI network Layer 7 that balances the workload, as well as providing certain security capabilities. The traffic to targets can be routed to compute instances, containers, IP addresses, and serverless functions based on the content of the request. As detailed in the following diagram, an ALB provides advanced request routing targeted at the delivery of modern application architectures, including microservices and container-based applications. An ALB improves security for the resources by ensuring that the latest SSL/TLS ciphers and protocols are used at all times.

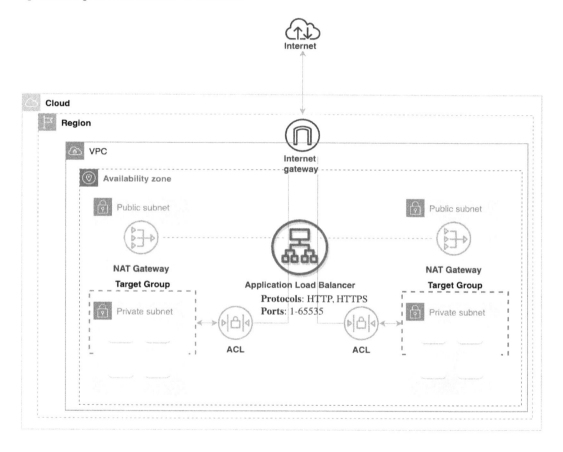

Figure 6.9 – Network protection – ALBs

As a best practice, limit the number of exposed applications to the internet. By default, apps and services that run within the cluster are not reachable over the public internet. When you keep your apps and services private, you can leverage the built-in security features to assure secured communication between resources. To expose services and applications to the public internet, look to leverage the network load balancer and ALB to support securely making the services publicly available.

In the case of modern deployments leveraging containers in a Kubernetes environment, similar patterns and principles apply. The ingress service offers TLS termination at two points in the traffic flow. The ingress ALB decrypts the package on arrival and load balances and forwards the HTTP/ HTTPS traffic to the cluster. For HTTPS traffic, the certificates need to be configured at the ALB level to manage the TLS termination. The ALB decrypts HTTPS requests before forwarding traffic to the final destination or downstream applications. If the downstream receivers require HTTPS, then the traffic is encrypted again and the certificates need to be configured at the target before it can be forwarded.

In a Kubernetes environment, where the ALB functions are available to be deployed on Pods, we can provide another level of network separation. This is done by creating edge worker nodes between the public network and a pool of compute worker nodes. ALBs are deployed as Pods to these edge worker nodes only. By deploying only ALBs to edge nodes, all Layer 7 proxy management is kept separate from the gateway worker nodes so that TLS termination and HTTP request routing are completed by the ALBs on the private network only.

For full protection of the cloud-hosted applications, a **web application firewall** (**WAF**) is required. WAFs provide an application-level IDS/IPS by providing detailed introspection and analysis of the application-level traffic. The following diagram shows how this capability can be deployed in front of web applications hosted on the cloud:

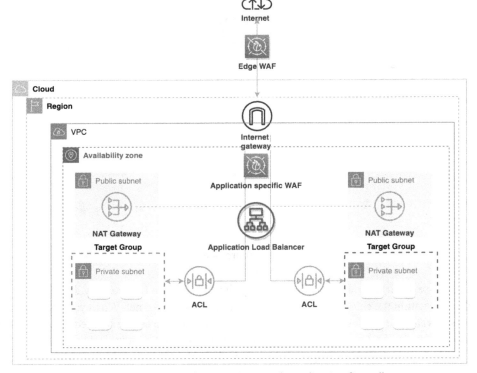

Figure 6.10 – Network protection – web application firewalls

This device protects a web application by detecting and blocking application-level attacks such as SQL injections and cross-site scripting. Using this additional protection layer in the cloud prevents threats without inhibiting traffic flow, data availability, and uptime. It helps you prepare for an unexpected data breach, outage, or disaster and know how to react to multiple, sustained attacks. A web application firewall can also block DDoS attacks through the following:

- Offering inline DDoS protection at a massive scale for network and application layers with real-time threat visibility

- Detecting and responding to parallel attacks with around-the-clock monitoring, while blocking web attacks such as SQL injections and cross-site scripting

- Routing traffic away from an infrastructure during an attack to help you better manage website availability and performance

WAFs can be deployed anywhere on the internet and configured to protect a specific DNS domain. Cloud-hosted firewalls enforce settings to protect your environment at the internet-cloud border and the cloud-on-premises border. This latter border is equally well protected by using an on-premises-hosted firewall.

We have seen several modes of isolation, connectivity, and protection in the hybrid world. As we have seen, traditionally, security has been enforced at the perimeter level. With mobile and cloud technologies exploding to edge infrastructures as well, securing the perimeter is no longer a viable option. Applications, users, and devices are most often outside of the trusted network perimeter that is traditionally protected by a security stack. With emerging 5G networks, a new era of interconnectivity is becoming a reality with millions of new devices connecting to the **Internet of Things** (**IoT**). So, the protected perimeter is now vanishing. The era of the trusted or protected network boundary is over.

So, to protect modern digital enterprises leveraging emerging technologies and built on a hybrid cloud, the thinking needs to be reversed. The model, as detailed in the following diagram, is called a Zero Trust network security model where nothing is trusted within or outside of the network perimeter. So, all network access requires strict identity verification for every person and device trying to access resources.

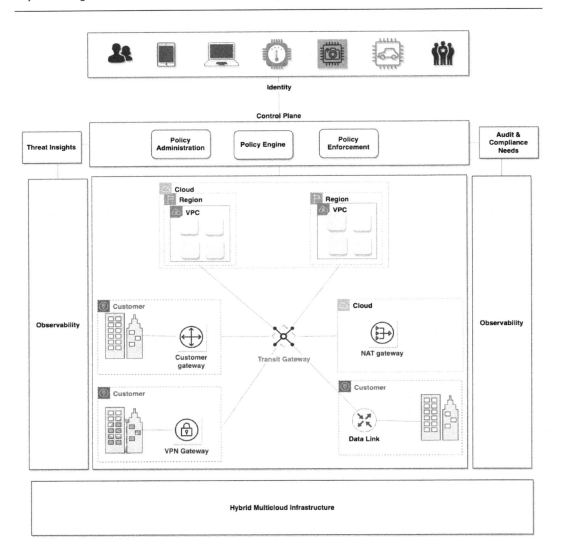

Figure 6.11 – Zero Trust network security pattern

The Zero Trust network security pattern as shown in the preceding figure is a holistic approach to network security that incorporates several different principles and technologies. The Zero Trust network security pattern draws on technologies such as multi-factor authentication, IAM, orchestration, analytics, encryption, scoring, and filesystem permissions. Zero Trust also calls for governance policies such as giving users the least amount of access they need to accomplish a specific task. The pattern shown in the preceding diagram is an emerging model based on zero trust. An enterprise cloud interconnect pattern would bring in an identity and context authorization control plane to determine the eligibility to connect to a subnet. Importantly in this model, depending on the micro-segmentation and security zones design, network security needs to determine which systems will have access to the resources across the environments or to put a firewall with centrally controlled policies between the major cloud interconnects. Many such policies and implementations have to be dynamic with the study of network traffic. This zero-trust software-defined network segmentation and connectivity model combines the network, identity, and contextual filtering across a hybrid cloud environment to determine access to a resource. The risk is calculated based on these dimensions, and based on the policies set in the control plane, the solution may or may not provide access to the network security zone.

Known uses

The following are the known uses:

- AWS Network Firewall is a managed service that protects Amazon VPCs. It has a flexible rules engine that lets you define firewall rules for fine-grained control over network traffic (https://aws.amazon.com/network-firewall).

- IBM Cloud's standard hardware firewall serves as a single firewall in a shared or multitenant mode. IBM also provides dedicated hardware firewalls.

- Using cloud-based service providers, such as Cloudflare and Akamai, provides a multilayered, robust approach to web protection so that customers can minimize the impact of these types of attacks on their web presence.

- Dedicated FortiGate-based security appliances provide enhanced security services such as stateful packet inspection, VLAN protection, ingress firewall rules, egress firewall rules, NAT, SSL VPN termination, IPsec VPN termination, advanced logging, and optional high-availability configuration, which can be inserted in the network in IBM Cloud.

- IBM Cloud Kubernetes Service is compatible with all IBM Cloud firewall offerings. For example, customers can set up a firewall with custom network policies to provide dedicated network security for their standard cluster and to detect and remediate network intrusion.

Summary

In this chapter, we learned how to isolate network traffic based on the purpose and target environments. We also learned how to establish secure connectivity in hybrid multi-cloud environments. Finally, we looked at the patterns to implement network protection for cloud workloads. The critical observation is how things are changing from a perimeter-based to a zero-trust-based model for network security.

In the next chapter, we will learn about data security patterns.

References

Refer to the following resources for more details on the topics covered in this chapter:

- Virtual Network – VPC | Microsoft Azure: `https://azure.microsoft.com/en-in/services/virtual-network/#overview`

- Implement network segmentation patterns – Microsoft Azure Well-Architected Framework | Microsoft Docs: `https://docs.microsoft.com/en-us/azure/architecture/framework/security/design-network-segmentation`

- What is Microsegmentation? – Palo Alto Networks: `https://www.paloaltonetworks.com/cyberpedia/what-is-microsegmentation`

- Internetwork traffic privacy in Amazon VPC – Amazon Virtual Private Cloud: `https://docs.aws.amazon.com/vpc/latest/userguide/VPC_Security.html`

- Advanced networking for IBM Cloud VPC – IBM Cloud Architecture Center: `https://www.ibm.com/cloud/architecture/content/course/advanced-networking-for-vpc/vpc-vs-classic-infrastructure/`

- Understanding Amazon VPC from a VMware NSX Engineer's Perspective | **AWS Partner Network (APN)** blog: `https://aws.amazon.com/blogs/apn/understanding-amazon-vpc-from-a-vmware-nsx-engineers-perspective/`

- Logically isolated VPC – Amazon VPC – AWS: `https://aws.amazon.com/vpc/`

- NAT gateways – Amazon VPC: `https://docs.aws.amazon.com/vpc/latest/userguide/vpc-nat-gateway.html`

- Differentiating between Azure Virtual Network (VNet) and AWS Virtual Private Cloud (VPC) – Developer Support: https://devblogs.microsoft.com/premier-developer/differentiating-between-azure-virtual-network-vnet-and-aws-virtual-private-cloud-vpc/

- One to Many: Evolving VPC Design | AWS Architecture Blog: https://aws.amazon.com/blogs/architecture/one-to-many-evolving-vpc-design/

- AWS VGW vs DGW vs TGW | Megaport: https://www.megaport.com/blog/aws-vgw-vs-dgw-vs-tgw/

- Zero Trust Architecture: A Brief Introduction – SSL.com: https://www.ssl.com/blogs/zero-trust-architecture-a-brief-introduction/

- Zero Trust for Government Networks: 6 Steps You Need to Know – Cisco Blogs: https://blogs.cisco.com/government/zero-trust-for-government-networks-6-steps-you-need-to-know

- IDS/IPS on AWS — Part I — An overview of available solutions by Paolo Latella | Towards AWS: https://towardsaws.com/ids-ips-on-aws-part-i-an-overview-of-available-solutions-2896ee7c6d27

- Gateway Load Balancer – AWS: https://aws.amazon.com/elasticloadbalancing/gateway-load-balancer/

- Traditional Azure networking topology – Cloud Adoption Framework | Microsoft Docs: https://docs.microsoft.com/en-us/azure/cloud-adoption-framework/ready/azure-best-practices/traditional-azure-networking-topology

- Hub-spoke network topology in Azure – Azure Architecture Center | Microsoft Docs: https://docs.microsoft.com/en-us/azure/architecture/reference-architectures/hybrid-networking/hub-spoke

- Azure best practices for network security – Microsoft Azure | Microsoft Docs: https://docs.microsoft.com/en-us/azure/security/fundamentals/network-best-practices

- Azure DDoS Protection features | Microsoft Docs: https://docs.microsoft.com/en-us/azure/ddos-protection/ddos-protection-overview

Part 4:
Data and Application
Security Patterns

Data protection is a critical requirement when leveraging data and storage resources in multiple clouds. This part will cover patterns for protecting data at rest, in transit, and in use. Key management and certificate management are other key services used in the context of data protection. We will look at the emerging patterns of how these services can be leveraged effectively to allow enterprises to keep full control of their data in a shared responsibility model of cloud data protection. Threat modeling involves understanding the threats and attacks on an application that can lead to security incidents. Secure engineering ensures that products, applications, and services are built with strong security and privacy controls. Configuration and vulnerability management are other critical capabilities needed for security and compliance programs. This part discusses the automation aspects of how to shift left and integrate security in the early stages of DevOps to create DevSecOps to ensure the application runs safely on hybrid multi-cloud environments.

This part comprises the following chapter:

- *Chapter 7, Data Security Patterns*
- *Chapter 8, Shift Left Security for DevOps*

7
Data Security Patterns

In this chapter, we will learn about the patterns that can be leveraged to secure data and storage resources in multiple clouds. The key areas covered will be as follows:

- Protecting data at rest
- Protecting data in transit
- Protecting data in use

The primary object of the data security patterns shared in the following diagram is to protect the data with respect to the following aspects:

- **Confidentiality** – Making sure that only the people who need to know are made aware of the information and that no one else can access it in an unauthorized way.

- **Integrity** – Ensuring that the content, while it is transferred or communicated, is not altered in transmission. Data integrity refers to maintaining and assuring the accuracy and consistency of data over its entire life cycle.

- **Availability** – Ensuring the data is available to the right authorized entities at the right time and is accessible.

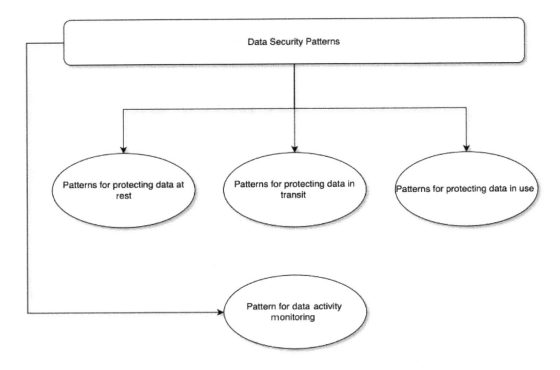

Figure 7.1 – Data security patterns

Patterns for protecting data at rest

Let's get started!

Problem

How to protect data at rest – files, objects, and data physically stored in storage services or databases.

Context

Data security is a critical element of hybrid multi-cloud security. There is exponential growth in the amount of data stored in multiple clouds and there is a need to protect this data. With increasing security threats, mounting regulatory requirements, and the explosive growth of data, effective data protection is more important than ever. Enterprises need to have systems in place to protect data— whether at rest or in transit—to protect from unauthorized access or exposure. Data subjected to industry-specific regulations and government regulations also requires specific considerations related to storage, protection, management, and governance of the data. The cloud provider needs to provide a mechanism that ensures the consumer's data is protected at rest. The pattern will discuss how to use encryption and key management patterns to protect data at rest.

Solution

Before we discuss the cloud data at rest protection patterns, let's understand foundation patterns that have been leveraged for data protection for several years.

Cryptography is the fundamental security technique leveraged to protect data at rest or in transit. Typically, cryptography involves converting plaintext into ciphertext (scrambled or hidden data) using algorithms. Essentially, if the data reaches unauthorized hands, they should not be able to make use of the information. Encryption is the process of converting a message into ciphertext and making the data unreadable. Decryption is the reverse process of recovering the clear data from the ciphertext. These algorithms provide a set of instructions on how to hide the data at the source by the sender and how to decrypt the data at the receiver end.

Cryptographic encryption and decryption are leveraged to protect the confidentiality and integrity of the data stored in on-premises and cloud-based systems. Encryption leverages keys for converting plaintext into ciphertext. This is typically done in two ways – symmetric and asymmetric encryption.

Symmetric encryption, as shown in the following figure, employs the same key for both encryption and decryption. In this model, the sender and receiver share a single common key. This same cryptographic key is used for both the encryption of plaintext and the decryption of ciphertext.

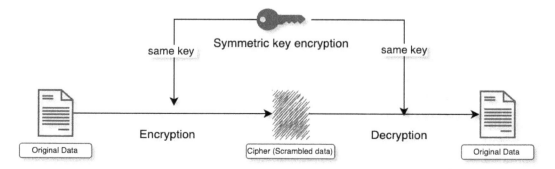

Figure 7.2 – Symmetric key encryption

In asymmetric key cryptography, as shown in the following figure, two different keys are used. One key is used for encryption and another key is used for decryption. These keys are typically issued by a **certificate authority** (**CA**). As shown in the following diagram, in this model, data is encrypted by its own public or private key. The data owner then shares the public key with everyone. Whoever wants to send data securely to the owner can encrypt it with the public key. Private keys are never shared by the data owner.

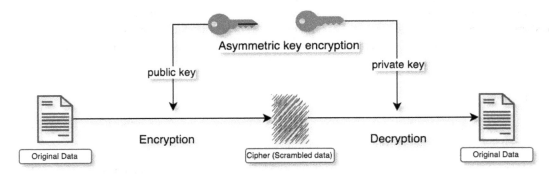

Figure 7.3 – Asymmetric key encryption

Any data encrypted for the data owner by senders with the public key can be decrypted by the owner with the private key.

In the shared responsibility model for the cloud, the way to do that is to provide control to the customer of the process or functionality by leveraging a **key management service** (**KMS**). This is a system or service that creates (and life cycle manages) the encryption keys. In the cloud world, maintaining and managing the keys is more critical than the encryption process for the security of the data. There is a need to control the access and use of the encryption keys. The security of the keys needs to be ensured and the responsibilities are split across the cloud provider and consumer. Key management services help clients manage the keys from unauthorized access or inadvertent access by unauthorized employee, thus meeting compliance requirements as well as making the sensitive data safe. To cryptographically protect data at rest, enterprises use encryption keys for specific data resources or instances such as object storage, files, or databases. To achieve these objectives, the pattern used is called the **envelope encryption** pattern.

In an envelope encryption pattern, as described in the following figure, data is encrypted with a **data encryption key** (**DEK**). The DEK is then further encrypted with a root key. Root keys help to protect DEKs stored in the cloud. The cloud consumer can wrap one or more DEKs with advanced encryption using the root key stored in the KMS that they can fully manage in the cloud. To achieve confidentiality and integrity, the root keys are typically stored in **hardware security modules** (**HSMs**), which ensures that only authorized people have access to these keys. The keys are created and managed with advanced encryption standards and algorithms within a trusted boundary. The root keys never leave the hardware device or the service boundaries. The root keys are used to protect the DEKs using the wrap/unwrap methods. The wrap/unwrap keys are also referred to as the **key-encryption key (KEK)**. Further enhancements to the security are done by separating the storage and access of DEK and KEK. For anyone to get back or decrypt the original data, they will need access to these keys managed by the key management service. To strengthen security, there can be a hierarchy of keys in this model.

These root keys are symmetric key-wrapping keys that are managed in the key management service. Further, the key management service integrates with the **cloud identity and access management (Cloud IAM)** to control fine-grained access to these keys. This is typically done by assigning advanced permissions and IAM roles to the user or service that needs access to the keys.

Any service or user that needs to encrypt the data needs to make an API call to the key management service to get a wrapped DEK. The service can specify the ID for the root key to be used for the wrapping function. The following diagram shows the wrapping operation details that provide confidentiality and integrity protection for the DEK. The DEK can also be generated, stored, and managed inside the key management service.

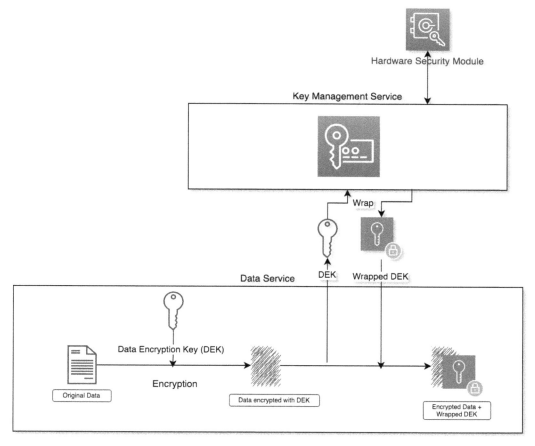

Figure 7.4 – Wrap function

In the unwrapping process, the key management service does the reverse of wrapping to provide the unwrapped DEK using the same algorithm. The requesting service or application sends a request to the key management service or API to unwrap the DEK. The details are shown in the following diagram. The request is made with the ID of the root key and the ciphertext value returned during the initial wrap request. The response to this request provides the unwrapped DEK. The DEK is then used to decrypt to return the original data or plaintext value to the requestor or service. Any additional security requirements can also be enforced during these operations.

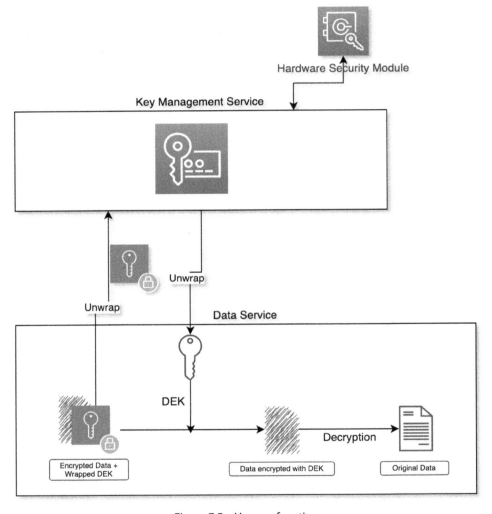

Figure 7.5 – Unwrap function

Envelope encryption thus combines the strength of key management and encryption algorithms to protect sensitive data in the cloud. This model also helps in the efficient rotation of encryption keys. Rotating keys on a regular basis helps meet regulatory requirements and industry standards, and is compliant with the best practices. Key rotation is done by retiring the current key and generating a new key to do rewrapping. With rotation at regular intervals, the lifetime of the key is reduced, which in turn reduces the probability of a security breach.

This method provides full control of their data for cloud consumers. The data services, such as object storage and database services, are integrated with the key management service. At any time, if the cloud consumers feel their key is compromised or detect any issue, they can mitigate the risk by rotating the keys. This pattern provides an efficient and cost-effective model to secure the cloud data services at scale and empower the customers to have more control of their data through key management.

Various patterns have emerged based on who manages the encryption keys as well as where the encryption is performed. The details are shown in the following diagram:

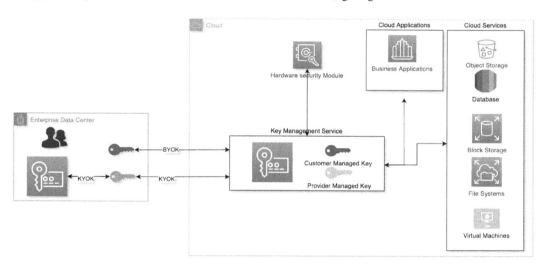

Figure 7.6 – Customer-managed encryption keys

More details are as follows:

- **Customer-managed encryption** – This is the pattern where the enterprise wants to keep complete ownership of the data. The regulation or business requirements mandate the encryption to be done on-premises, which provides the highest assurance for protecting their data at rest. This happens typically on the client side of the application and in a customer-controlled environment. The encryption may be performed using customer-managed keys or provider-managed keys.

- **Provider-managed encryption** – In this model of cloud provider-managed encryption, it employs encryption processing in the provider-controlled environment. Cloud provider-managed encryption always is performed on the server side. For instance, in the case of a cloud-based data service, when new data is received, it is encrypted before it is written to disk. Similarly, when data is requested by a client, the data is decrypted in the provider environment and made available to the requestor.

Customer-managed encryption provides the most control for protecting stored data, while the provider-managed approach simplifies the encryption process. Similarly, depending on who manages the encryption keys, the models supported are given as follows:

- **Cloud provider-managed keys** – In this pattern, the data at rest is protected by cloud provider-managed encryption with cloud provider-managed keys. The cloud provider does the data encryption with keys generated on the server side and stores it on the disk. The encryption process and keys are both managed by the provider. When access to the data is requested, it is decrypted and provided to the consumer transparently. This pattern provides isolation of encryption and key management ensuring the protection of the data at rest from physical theft or copying from the storage.

- **Customer-managed keys** – In this pattern, the root keys are provisioned and managed in the HSM. Customers can get dedicated HSMs from the cloud or leverage shared services to keep their keys. A variation of this is the **Bring Your Own Keys (BYOK)** option where the customers can import the existing root keys created elsewhere into the key management service on the cloud. Customer-managed encryption works by isolating all cryptographic processing to a customer-controlled environment. Customers have the flexibility to create and manage their own encryption keys on the cloud. They can assign different keys for different data resources and encrypt or decrypt sensitive data. The keys can be life cycle-managed by using an on-premises key management system or cloud key management service. All cryptographic operations take place in an environment that is fully under the control of the customer. The cloud provider cannot access any **data in the clear** (other term for **unencrypted data**). The provider does not also have access to the encryption keys that protect that data. This model is suitable for running highly sensitive workloads on the cloud with complete control of security with the customers.

The decision on where the encryption process should be run and who should manage the encryption keys is based on the workload and data characteristics. For a multi-cloud solution that requires tight data protection at rest, customer-managed encryption with customer-managed encryption keys is recommended with dedicated HSMs. Different data storage types can be protected at rest with these patterns individually such as databases, block storage files, and objects. The pattern can also be applied to protect the individual resources managed by the services such as object storage services and data services.

Encryption as a service is an emerging model in the cloud that is described in the following figure. In this hosted service model, you can spin up the service on the cloud and there is no need to manage and maintain HSMs. This is recommended for protecting highly sensitive data against breaches caused by attacks or unauthorized access from internal users. The encryption and decryption happen in a highly secured compute environment preventing any data tampering.

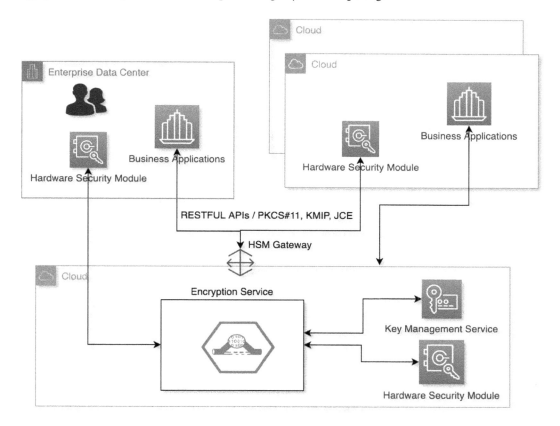

Figure 7.7 – Encryption as a service pattern

An extension to this pattern for the multi-cloud world is the multi-cloud key orchestrator pattern. In this pattern, a central key management service can integrate with a hybrid deployment across public, private, and on-premises data centers to protect the data at rest supporting **Keep Your Own Key (KYOK)** as well BYOK patterns.

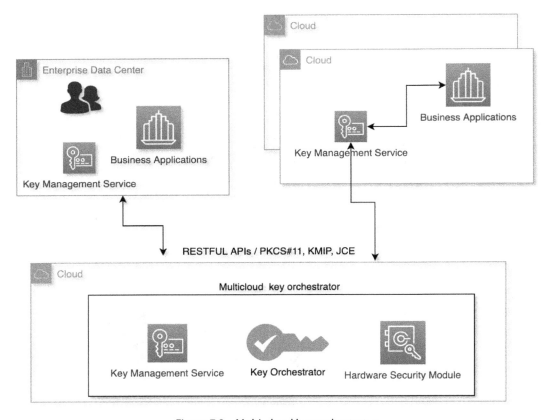

Figure 7.8 – Multi-cloud key orchestrator

Known uses

The following are the known uses:

- IBM Cloud supports provider-managed encryption keys as well as customer-managed encryption keys patterns. The root keys are the key-wrapping keys to protect the DEKs that are used to encrypt the data and storage services. **Advanced Encryption Standard (AES)** algorithms are used for encryption. The keys are generated within the trust boundary of IBM Cloud HSMs, so only the clients have access to your encryption keys. As a developer, you can choose to encrypt data at any of the layers in the data-storage stack using this pattern. You can choose to encrypt at the following layers:

- **Storage layer** – Protects against physical theft of storage drives. The storage layer encryption is transparent to layers above it in the stack and data is in clear for all layers above them. Does not provide any protection from users at the hypervisor/OS layer.

- **Hypervisor/OS layer** – Ensures that only the hypervisor/OS that has access to the encryption keys has access to the data. Storage volumes cannot be attached to another hypervisor/OS that does not have access to the encryption keys. Provides protection from unprivileged hypervisor users but still does not provide protection from hypervisor/OS admin users.

- **Database layer** – Protects data from storage and hypervisor/OS admins and provides complete customer control when the database is managed by the customer. Data in clear is exposed only to the database admins and applications that have access to the database.

- **Application layer** – Provides the highest level of security since data in clear is only available in the application where it is processed, and all the layers below have access to only encrypted data. This protects data from all system and database admins but needs considerable effort to redesign the application and database layers, as this can introduce database performance issues and encrypt key management issues.

- The encryption-as-a-service model is supported by a few vendors such as Fortanix, which can help protect the data across on-premises and cloud through a hosted service that does encryption and key management in a controlled environment.

- The multi-cloud key orchestrator pattern is implemented by IBM as the Universal Key Orchestrator, which helps enterprises manage KYOK and BYOK models for managing data encryption keys across multiple key stores across multiple cloud environments, including keys managed on-premises, on IBM Cloud, AWS, and Microsoft Azure (`https://www.ibm.com/cloud/hyper-protect-crypto`).

- **AWS Key Management Service** (**AWS KMS**) helps implement the data at rest protection patterns to create and manage keys and control the use of encryption across a wide range of AWS services and applications. AWS KMS is a secure and resilient service that uses FIPS 140-2 validated hardware security modules to protect your keys.

- All Azure-hosted services are committed to providing **encryption at rest** options. Azure services support either service-managed keys, customer-managed keys, or client-side encryption. Azure services are broadly enhancing encryption at rest availability.

- In the case of Google Cloud, data is chunked and encrypted with **DEKs**. DEKs are encrypted with **KEKs**. KEKs are stored in KMS that are run on multiple machines in data centers globally. KMS keys are wrapped with the KMS master key, which is stored in the root KMS. The root KMS is much smaller than KMS and runs only on dedicated machines in each data center. Root KMS keys are wrapped with the root KMS master key, which is stored in the root KMS master key distributor. The root KMS master key distributor is a peer-to-peer infrastructure running concurrently in RAM globally on dedicated machines; each gets its key material from other running instances. If all instances of the distributor were to go down (total shutdown), a master key is stored in (different) secure hardware in (physical) safes in limited Google locations.

Protecting data in transit patterns

Let's get started!

Problem

Data is exchanged between the requestor and receiver. How do we protect this data that is being exchanged or transmitted over the internet or intranet?

Context

The data exchanged or transmitted over the net is often referred to as **data in transit** or **data in motion**. In a hybrid multi-cloud world, data needs to be exchanged across several on-premise and cloud applications. There are several service-to-service communications that also happen on-premise or within the cloud that need to be protected. With digitization, the data is accessible from many devices by employees as well as users from customer and partner organizations. There are industry and geography-specific guidelines such as PCI DSS, GDPR, and HIPAA that mandate the data to be safeguarded meeting regulatory guidelines. Loss of data to the wrong hands can possibly incur financial penalties and reputational risks. Data in motion poses many risks and threats from both inside and outside the enterprises including user error, insecure networks, and man-in-the-middle attacks.

Solution

Encryption is used to protect the data in motion. Data should always be encrypted when it's moving through many internal or external networks. This includes encrypting all data prior to transport or through protected tunnels. HTTPS, SSL, and TLS are the terms often used in the context of the protection of data in motion. Let's understand these terms better. **HTTPS** is the secured version of **HyperText Transfer Protocol** (HTTP) and is the protocol used by the client-server application, such as a browser talking to web servers to exchange information. When that exchange of data is encrypted with SSL/TLS, then we call it **HTTPS**. The **S** stands for **Secure**. **SSL** stands for **Secure Sockets Layer**. IETF, which managed the release of SSL, later renamed it **TLS**, which stands for **Transport Layer Security**. Digital certificates form a critical part of implementing a security environment in which two parties communicate securely and send confidential messages to each other using TLS. To understand digital certificates, we need to understand a few basic concepts about public key cryptographic systems. In a public key cryptographic system, a pair of related cryptographic keys are used, one for encryption and the other for decryption. One key, the private key, is kept secret and securely in the possession of owner A. The other key, the public key, is provided to anybody wishing to communicate with owner A. The public key of a public-private key pair must be made available to other users. This public key is made available in a digitally-signed document called a **certificate**. A certificate, among other things, contains a user's name and a user's public key, digitally signed by a CA, which is an entity that issues digital certificates and is part of a certificate management system.

In organization-validated certificates, the certificate contains the verified name of the entity that controls the website for which the certificate has been issued. These kinds of certificates are used for websites conducting e-commerce or financial transactions. **Extended-validation** (**EV**) certificates are issued after even more extensive validation and provide a higher level of assurance than organization-validated certificates. EV certificates are recommended for external-facing endpoints that involve financial transactions.

Domain-validated certificates are getting popular with cloud adoption. The domain for which the certificate is issued is verified and the certificate is issued through automated processes. No check is made regarding the organization requesting the certificate. These certificates are best for internal-facing microservices of a multi-cloud application.

Other considerations for the type of certificate to request include the length of the asymmetric key pair generated for the certificate and the duration for which the certificate is valid. You should select 1,024-bit or 2,048-bit keys, with 2,048-bit keys being more secure and being the only option for extended validation certificates. Certificates should be set to expire within a year or two. Domain-validated certificates are set to expire and rotate in 90 days.

During a TLS/SSL handshake, a public-key algorithm, usually RSA, is used to securely exchange digital signatures and encryption keys between a client and a server. This identity and key information is used to establish a secure connection for the session between the client and the server. After the secure session is established, data transmission between the client and server is encrypted using a symmetric algorithm, such as AES. The client requests a TLS/SSL connection and lists its supported cipher suites. The server responds with a selected cipher suite. The server sends its digital certificate to the client. The client verifies the validity of the server certificate, for authentication purposes. It can do this by checking with the trusted CA that issued the server certificate or by checking in its own key database. The client and server securely negotiate a session key and a **message authentication code** (**MAC**). The client and server securely exchange information using the key and MAC selected.

Modern age emerging trends point to increased instances using certificate-based authentication such as microservices-based hybrid applications, connectivity for **Internet of Things** (**IoT**)-enabled services, machine-to-machine transactions, and so on. Protecting data in motion then becomes more about managing the certificates and ensuring control of the private keys used for TLS including limiting access and having traceability. This is facilitated at a high level, with a certificate management system that provides services for automated certificate management. This includes managing, issuing, and revoking digital certificates for various entities such as users, websites, applications, and devices such as servers and mobile devices. Using a cloud-based certificate management system allows businesses to focus on developing their core business applications. Certificate management needs to prevent outages by making sure certificates are renewed on time and deployed to all the right places.

As described in the following diagram, data in transit is protected using the pattern to keep encrypted end to end between the communicating parties. HTTPS and TLS/SSL-based communication is ensured across all internal and external services. Certificate management is leveraged for creating and deploying certificates to the communicating endpoints. Certificate management also does the life cycle management activities such as revoking, rotating, or retiring certificates. Certificate management can be leveraged to push certificates to the communicating endpoints such as service ingresses, load balancers, content delivery networks, virtual machines running services, and isolated execution environments.

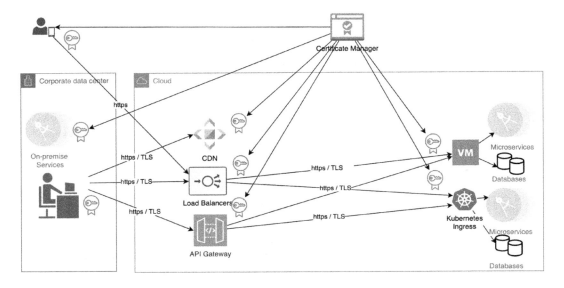

Figure 7.9 – Data in transit protection and certificate management

Another solution that is becoming more popular in protecting data when it is moved to the cloud is data tokenization. Encryption, as we learned, is the process of converting a message into ciphertext and making the data unreadable using cryptographic keys. In tokenization, the original data is replaced by a random value called the token that is completely unrelated. Tokenization is typically done using mathematical transformation. Enterprises need to hold sensitive PII such as social security numbers or credit card data within their systems. This may include data such as card numbers and expiry dates that are stored by enterprises to make online transactions convenient for their customers. But saving this type of crucial information in multiple enterprise systems increases the risk multifold for the business user. Using tokenization, organizations can continue to use the data for analysis and business purposes without incurring the risk of storing sensitive data internally. The token is used in the business systems in the same ways as the original data. Data security is achieved by storing the original data in a secure cloud token vault, along with the encryption and tokenization keys. Importantly, the responsibility of protecting the data shifts from the enterprise to the organization owning the token vault. Security for the token vault is better managed by the regulatory body with tighter access controls.

In this pattern, as shown in the following diagram (steps a and b), if an enterprise does not want to share critical and confidential information with the applications running on the cloud, it can tokenize the data before pushing it to the cloud. The data will be very similar in syntax and semantics so that the business application will not have any impact on executing its functions. The business systems treat the token in the same way as the original data.

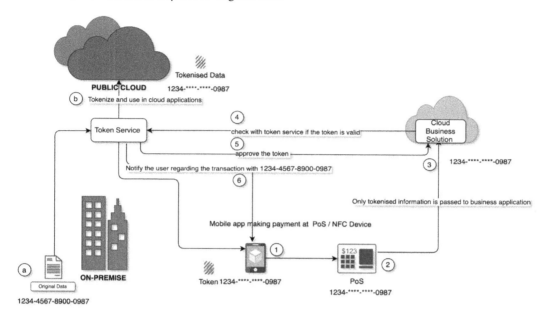

Figure 7.10 – Data tokenization

Another example scenario of this pattern in action is the use case of a mobile application or wallet that needs to store credit card information for online transaction convenience. In this case, the mobile application tokenizes the card information and only stores the token on the client's device. When a transaction is performed at the point of sale machine or through NFC devices, the token is exchanged with the merchant application. The application can then check the validity of the token with the token service or vault. The permitted card network maintaining the token service can keep the information secure in a single place and can then notify the user regarding the transaction related to the card with the owner through the mobile application.

Known uses

The known uses are as follows:

- AWS offers several options including load balancing services (for example, Elastic Load Balancing, Network Load Balancer, and Application Load Balancer), Amazon CloudFront (a content delivery network), and Amazon API Gateway for terminating network connections. In order to implement a secure (TLS) connection, each of these endpoint services allows customers to upload their own digital certificates to bind a cryptographic identity to the endpoint. AWS simplifies the process of generating, distributing, and rotating digital certificates with **AWS Certificate Manager (ACM)**.

- Microsoft gives customers the ability to use the TLS protocol to protect data when it's traveling between the cloud services and customers. Microsoft data centers negotiate a TLS connection with client systems that connect to Azure services. TLS provides strong authentication, message privacy, and integrity (enabling detection of message tampering, interception, and forgery), interoperability, algorithm flexibility, and ease of deployment and use.

- IBM Certificate Manager helps manage and deploy SSL/TLS certificates to apps and services. Certificate Manager provides a security-rich repository for certificates and associated private keys. It can also alert you when your certificates are about to expire and prevent outages.

- Google encrypts and authenticates data in transit at one or more network layers when data moves outside the physical boundaries not controlled by Google or on behalf of Google. All VM-to-VM traffic within a VPC network and peered VPC networks is encrypted. Depending on the connection that is being made, Google applies default protections to data in transit. For example, we secure communications between the user and the **Google Front End (GFE)** using TLS. Google Cloud addresses additional requirements for encryption with IPSec tunnels, Gmail S/MIME, and managed SSL certificates.

- Istio – Service-to-service security is a key concern in the microservices world. Conventional network security approaches fail to address security threats to distributed applications deployed in dynamic cloud environments. Istio is an example of a service mesh that secures service-to-service communication tunneled through high-performance client-side and server-side (envoy) proxies. The communication between the proxies is secured using mutual TLS. The benefit of using mutual TLS is that the service identity is not expressed as a bearer token that can be stolen or replayed from another source. Istio authentication also introduces the concept of **secure naming** to protect from server spoofing attacks – the client-side proxy verifies that the authenticated server's service account is allowed to run the named service.

- In order to enhance the digital payment experience and add an extra layer of security, the **Reserve Bank of India** (**RBI**) has made it mandatory for all credit and debit card data used in online, point-of-sale, and in-app transactions to be replaced with unique tokens.

- TokenEx is a tokenization solution provider that captures sensitive data via API, batch, or third-party requests and then tokenizes that data to de-risk and remove downstream systems from the scope.

Data in use

Let's get started!

Problem

How do I protect data in use?

Context

Today, data is often encrypted at rest in storage and in transit across the network, but applications and the sensitive data they process are vulnerable to unauthorized access and tampering at runtime. Most governments and highly regulated industries that deal with sensitive and **Personal Identifiable Information** (**PII**) are concerned about protecting their data while in use. This is required from regulation requirements as well as from data privacy standards. This pattern discusses the way to leverage technologies and systems to protect the security and confidentiality of such customer data.

Solution

Confidential computing, as shown in the following diagram, protects data and applications by running them in secure enclaves that isolate the data and code to prevent unauthorized access, even when the compute infrastructure is compromised. While confidential computing is revolutionizing how customers protect their sensitive data, organizations need to simplify the process of creating enclaves, managing security policies, and enabling applications to take advantage of confidential computing.

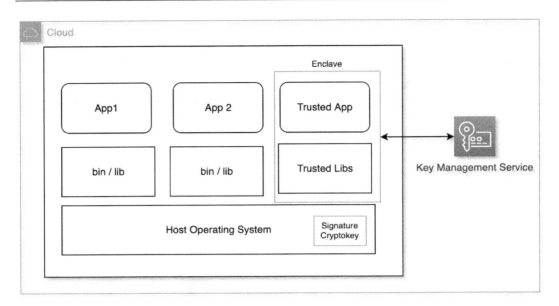

Figure 7.11 – Confidential computing

With the use of specialized hardware and associated firmware, enterprises can protect their code and data while it is loaded into the machine's memory. While external access is provided to the client to trigger and manage the execution of the process in the machine, confidential computing does not provide even the cloud provider with access to the data. So, even if the compute is compromised, the data is protected from any external access. Also, this pattern allows enterprise customers to confirm that their workloads are running on trusted components, and they cooperate across the layers to confirm that the confidentiality of the code and data is maintained. This pattern combines the leveraged trusted computing pattern that we learned about in *Chapter 5*. There are also cloud offerings that provide policy-based controls determined by the data owner and enforced in a way across the life cycle to provide security and privacy. There are data-in-use protection services that provide capabilities to provision, manage, maintain, and monitor multiple database types through standardized APIs. These are security-rich data stores that are provisioned on the cloud with in-memory protection and customers have full control of data through integration with the key management service.

Known uses

The known uses are as follows:

- Azure confidential computing encrypts data in memory in hardware-based trusted execution environments and processes it only after the cloud environment is verified, helping prevent data access by cloud providers, administrators, and users. This secures sensitive and regulated data while it's being processed in the cloud.

- IBM Cloud Data Shield enables users to run containerized applications in a secure enclave on an IBM Cloud Kubernetes Service host, providing data-in-use protection. IBM Cloud Data Shield supports user-level code to allocate private regions of memory, called **enclaves**, that are protected from processes running at higher privilege levels. Customers can run their workloads on SGX nodes running the Kubernetes-based IBM Container service. The pattern leverages enclaves provided by Intel SGX servers to protect data in memory.

- AWS Nitro System provides data-in-use protection requirements with Nitro-based **Amazon Elastic Compute Cloud** (**Amazon EC2**) instances, without requiring any code or workload changes from the customer side. AWS Nitro Enclaves provides a way for customers to use familiar toolsets and programming models to achieve data-in-use protection.

- Google Cloud customers can encrypt data in use, taking advantage of security technology offered by modern CPUs (for example, Secure Encrypted Virtualization extension supported by 2nd Gen AMD EPYC™ CPUs) together with confidential computing cloud services. This gives the confidence to customers that their data will stay private and encrypted even while being processed.

Data classification and monitoring patterns

Let's get started!

Problem

How do we discover the data and classify its existence to determine the protection strategy and mechanisms? The problem is also about solving how to monitor the access of data and alert on any important event or breach.

Context

Before determining the protection of data, enterprises need to acknowledge the existence of data. Then, it is required to classify the data based on the sensitivity to determine what data protection and monitoring strategy needs to be followed.

Many enterprises are unaware of their data distribution and sensitive content residing outside of expected security protocol. The other problem is that these organizations retain far too much data that may be redundant, obsolete, or trivial. Redundant data is data that is duplicated, and obsolete data has passed its useful life. Data that has no ongoing business value is considered trivial. Typical enterprises struggle with dark data that they have no insights into. This could be unstructured data that may be discovered during a data breach and stored in sources such as email, chat, file shares, and so on. Then there is sensitive data such as PII, **highly confidential information (HCI)**, and **payment card industry (PCI)** data that needs greater protection and assurance. Data privacy controls how information (particularly about individuals) is collected, used, shared, and disposed of, in accordance with policies or external laws and regulations. Enterprises need to plan data protection based on its sensitivity, location, storage, and handling required to meet the security and privacy measures for complying with the policy.

Solution

While data security relates to the protection of personal data against loss, unauthorized access, destruction, use, modification, disclosure, and so on, data privacy relates to the rights of individuals to control how their public and non-public personal data is collected and used. Data governance defines who can take what actions with what personal data, when, under which circumstances, and using what methods. This also includes the set of policies that defines how the enterprise manages its information throughout the data life cycle to reduce operational and systemic risk—without adversely impacting the value of the information.

For data security, data protection, and governance, the first step is data discovery and classification. Data discovery is the process of finding the existence of data in the enterprise locked in different sources. The data protection and security controls are determined based on the classification.

Typically, these steps are automated to provide a seamless approach for finding, classifying, and protecting critical data, whether in the cloud or on-premises. Data classification often involves categorizing the data discovered under a few defined types. The various levels of classification include the following:

- Public data is important information but is often available or freely accessible for people to read. This is typically the lowest level of data classification as it poses little or no risk to the enterprise. Press releases, enterprise information, and so on that are updated at periodic intervals and shared with a wider audience are all examples.

- Private data is information that's prudent to keep from public access to best protect the integrity of the information and access to other data through it. Private data is shared on a need-to-know basis with proper identification of the user requesting access to the information. Personal contact information, such as email addresses and phone numbers, and research data are examples of private data.

- Internal data often relates to an enterprise and only the employees typically have access to internal data. Some examples of internal data can include an enterprise's business plans and strategies, revenue projections, and so on. Internal data can have different levels of security and access according to the sensitivity of the information. For example, executives might have access to some internal data that is not made available to the junior members of the team.

- Confidential data is sensitive information made available only to a limited group of people with proper authentication and authorization. Some examples of confidential data include PII such as social security numbers, medical or health records, credit card details, or financial records. Confidential data shared outside the enterprise often involves signing a **non-disclosure agreement (NDA)** to further protect the integrity of the data.

- Restricted data is the most sensitive of the data classifications requiring strict security controls. It requires data encryption at all times to prevent malicious users from accessing or reading the content. Restricted data, if leaked, can pose an impact on the enterprise's business and reputation. It can also result in some hazards and raise safety issues for human or enterprise assets. Examples of restricted data can include protected health information, personal information, and business-critical information assets (for example, crown jewels).

Data classification and data activity monitoring are two effective methods to help secure critical information. Before we can adequately protect sensitive data, we must identify and classify its existence. Automating the discovery and classification process is a critical component of a data protection strategy to prevent a breach of sensitive data. As part of the automation, predefined data classes with managed labels can be used to classify information based on confidentiality and security requirements. There are tools also that do automatic profiling to detect data content, including sensitive and PII data. Thus, integrated data discovery and classification capabilities provides a seamless approach to finding, classifying, and protecting critical data, whether in the cloud or in the data center.

As described in the following diagram, data protection, monitoring, and governance policies are developed and applied based on the protection needed for sensitive data (for example, crown jewels):

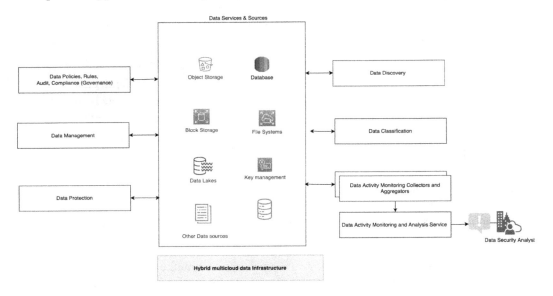

Figure 7.12 – Data activity monitoring

Operationalizing data protection needs to take a risk-based approach to create enterprise-wide, holistic policies covering data-in-use, data-in-transit, and data-at-rest. The governance and roles/responsibilities will need to be developed. For an enterprise, there are several data stores involved with different data velocities, variety, and veracity. Data governance needs visibility and control over this data spread across hybrid multi-cloud, data-in-motion, and data permeating enterprise boundaries. Data governance policies and rules are established to capture enterprise or corporate-level data mandates. Given that management of the encryption key is the most important aspect of cloud data protection, all access to encryption keys must be logged to provide full visibility and historical tracking. This should include all user actions that are performed on encryption keys.

Known uses

The following are the known uses:

- AWS recommends starting with a three-tiered data classification approach to sufficiently meet both public and commercial customer needs and requirements. For organizations that have more complex data environments or varied data types, secondary labeling is helpful without adding complexity with more tiers. With AWS, you can implement database activity streams to monitor and report activities. The stream of activity is collected and transmitted to a data streaming and analysis service such as Amazon Kinesis. From Kinesis, you can monitor the activity stream, or other services and applications can consume the activity stream for further analysis.

- The following is a classification that Microsoft uses but through the use of resource tags, customers can apply a data classification model to their deployed resources on the cloud (see the reference link in the *References* section for more details):

  - **Non-business**: Data from your personal life that doesn't belong to Microsoft

  - **Public**: Business data that is freely available and approved for public consumption

  - **General**: Business data that isn't meant for a public audience

  - **Confidential**: Business data that can cause harm to an organization if overshared

  - **Highly confidential**: Business data that would cause extensive harm to an organization if overshared

- Azure Monitor is a solution that does collecting, analyzing, and acting on telemetry from the cloud and on-premises environments. This helps to proactively identify issues that affect the resources and trigger actions such as analysis, alerting, and streaming to external systems.

- Google Cloud **Data Loss Prevention** (**DLP**) helps to understand, manage, and protect sensitive data. With the Cloud DLP, customers can easily classify and redact sensitive data contained in text-based content and images, including content stored in Google Cloud storage repositories.

- IBM Guardium supports the discovery and classification of sensitive data to allow the creation and enforcement of effective access policies.

Summary

In this chapter, we learned about the core patterns that can be leveraged to protect data at rest, in transit, and in use. We also learned how important a key management service is when it comes to protecting data at rest, and the significance of certificate management to protect data at rest. Finally, we looked at some of the emerging patterns such as the multi-cloud key orchestrator and encryption-as-a-service patterns. We also learned how and why it is important to implement a data activity monitoring pattern to measure the effectiveness of the data protection mechanisms and provide alerts on any threats. The critical observation is how, in a shared responsibility model of the cloud, customers can leverage key management services to be in full control of their data.

In the next chapter, we will look at shift-left security for DevOps.

References

Refer to the following for more information about the topics covered in this chapter:

- Data at rest and transit protection patterns supported by AWS: `https://docs.aws.amazon.com/whitepapers/latest/logical-separation/encrypting-data-at-rest-and--in-transit.html`

- Data at rest protection: `https://wa.aws.amazon.com/wat.question.SEC_8.en.html`

- Data at rest and transit protect patterns supported by IBM Cloud: `https://www.ibm.com/cloud/architecture/architecture/practices/data-security/`

- Data security architecture: `https://www.ibm.com/cloud/architecture/architectures/data-security-arch/reference-architecture`

- IBM Unified Key Orchestrator: `https://www.ibm.com/cloud/blog/announcements/unified-key-orchestrator`

- Azure data at rest protection patterns: `https://docs.microsoft.com/en-us/azure/security/fundamentals/encryption-atrest`

- Azure data in transit protection patterns: `https://docs.microsoft.com/en-us/azure/security/fundamentals/encryption-overview#encryption-of-data-in-transit`

- Google cloud encryption at rest: `https://cloud.google.com/docs/security/encryption/default-encryption`

- Google cloud data in transit protection: `https://cloud.google.com/docs/security/encryption-in-transit`

- Istio security: `https://istio.io/latest/blog/2017/0.1-auth/`

- Customers can combine Fortanix with Intel SGX to ensure neither app nor data is ever decrypted in any insecure state: `https://www.fortanix.com/solutions/use-case/confidential-computing`

- IBM Data Shield: `https://www.ibm.com/cloud/data-shield`

- Azure confidential computing: `https://azure.microsoft.com/en-us/solutions/confidential-compute/#overview`

- AWS confidential computing: `https://aws.amazon.com/blogs/security/confidential-computing-an-aws-perspective/`

- Google confidential computing: `https://cloud.google.com/confidential-computing`

- AWS data classification recommendations: `https://docs.aws.amazon.com/whitepapers/latest/data-classification/data-classification.html`

- Microsoft data classification: `https://learn.microsoft.com/en-us/azure/cloud-adoption-framework/govern/policy-compliance/data-classification`

- Azure Monitor: `https://docs.microsoft.com/en-us/azure/azure-monitor/overview`

- Google classification, redaction, and de-identification: `https://cloud.google.com/dlp/docs/classification-redaction`

- IBM data activity monitoring: `https://www.ibm.com/docs/en/guardium`

- AWS guidance to determine your requirements for tokenization, with an emphasis on the compliance lens: `https://aws.amazon.com/blogs/security/how-to-use-tokenization-to-improve-data-security-and-reduce-audit-scope/`

- *Meet PCI compliance with credit card tokenization*: `https://azure.microsoft.com/en-us/blog/meet-pci-compliance-with-credit-card-tokenization/`

- Protect sensitive data by using tokens: `https://www.ibm.com/cloud/architecture/architectures/security-data-tokenization-solution/`

- *Take charge of your data: How tokenization makes data usable without sacrificing privacy*: `https://cloud.google.com/blog/products/identity-security/take-charge-of-your-data-how-tokenization-makes-data-usable-without-sacrificing-privacy`

- RBI Notification on credit card: `https://rbi.org.in/Scripts/NotificationUser.aspx?Id=11449&Mode=0`

- Tokenex data tokenization solution: `https://www.tokenex.com/solutions/data-protection`

Shift Left Security for DevOps

This chapter will discuss the different patterns we can use to infuse security into the DevOps pipeline. **Shift left security** is required from the first stages of concept, development, and operations to ensure the application runs safely in hybrid multi-cloud environments. Threat and vulnerability management are critical aspects of security and compliance programs. Enterprises are incorporating security into their DevOps pipelines to create DevSecOps pipelines. This chapter will discuss the patterns for identifying vulnerabilities in the cloud resources across infrastructure, middleware, and applications and how to remediate them. Configuration management is another important topic we will cover, which specifies how to manage and control configurations for cloud resources to enable security and facilitate the management of risk.

In this chapter, we will cover the following topics:

- Secure engineering and threat modeling
- The DevSecOps pattern

The following diagram illustrates the various shift left security patterns:

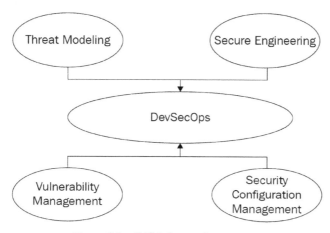

Figure 8.1 – Shift left security patterns

We will be covering all of these different aspects in this chapter.

Secure engineering and threat modeling

Let's get started!

Problem

How can we incorporate security early in the application's conceptualization and development cycle?

Context

In the current scenario, enterprises need to continuously incorporate customer feedback and deliver new capabilities faster. In cloud-native development, applications are developed in an agile model with **continuous integration** (**CI**) and **continuous delivery** (**CD**). Traditionally security was considered an afterthought or referred to as bolt-on security. Unlike traditional approaches for meeting the demands of the modern world, development and operations need to come together as DevOps. As described in *Figure 8.2*, the model is often referred to as shift left security, which essentially means considering security early in this approach during the stages of conceptualization, design, development, and deployment – that is, DevSecOps. This will ensure that the application is securely developed and is free from vulnerabilities and can be safely operated.

Solution

To create a secure cloud application, security needs to be integrated into all the steps of the application development cycle. This would start with secure engineering, which essentially looks at the entire stack in terms of the infrastructure, platform, and application. In this pattern, we will focus on and understand the threats and risks related to the application design, development, and deployment. Enterprises define and follow a secure engineering framework that looks at security across architecture, design, and development processes. When the application is consuming resources across data centers and multi-cloud environments, ensuring the security of the integrated supply chain is also an important dimension of the secure engineering model.

Secure engineering is the first step in that direction to ensure the products and services are built with strong security and privacy controls from the beginning, which helps them comply with industry regulations and compliance standards.

Some of the important considerations during the different phases of cloud application development are as follows:

- Who are the actual users of the product?
- What sensitive, **personally identifiable information** (**PII**), or client information is collected?

- Does the application meet the expected industry and regulatory standards for the user community it is serving?

- Where is the application deployed, what resources are being used, how is the application operated, and who has access to the data?

As part of secure engineering, we also look at the extent of damage that a bad actor in the application or system can do. This should include the actors accessing IT infrastructure, data, applications, and APIs from the **Internet of Things (IoT)**/operational technology layer as well. This leads us to the topic of threat modeling and management:

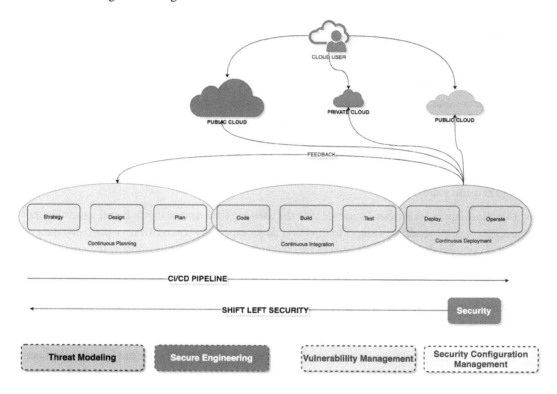

Figure 8.2 – Shift left security

Threat modeling involves understanding the threats and attacks on an application that can lead to security incidents. Threat modeling covers all parts of the stack, including the infrastructure (compute, storage, and network), platform, and application services and components. In this chapter, we will look more closely at the application threats. This will start with getting a deeper understanding of the users of the application, their access patterns, and identifying vulnerabilities in the process. Threat modeling extends to developing code and deploying the application and its operations.

Threat modeling lists the set of possible conditions and consequences that can lead to incidents, resulting in loss or exposure of sensitive data. This may be done as a simple set of practices integrated with the development process. A simple sanity check for application security will include things such as auditing the user authentication and authorization models, as well as checking for common application programming errors that provide bad actors to attack the application. The *OWASP Top 10* is a project that captures and categorizes critical security risks to applications. At a minimum, checks should be in place to verify that the application addresses these risks. The top 10 application security risks based on 2021 data are as follows:

- **Broken Access Control**: This risk is primarily seen alongside the violation of the principle of least privilege or deny-by-default policies, which leads to unauthorized information disclosure or changes. The attacker typically gains access by accessing the UI or API through missing access controls or elevation of privilege as an admin user while logged in as a normal user. Most cloud applications rely on **JSON Web Token (JWT)** access control and attackers manipulate JWTs to gain elevated privileges. CORS misconfiguration is another broken control that allows API access from unauthorized/untrusted origins.

- **Cryptographic failure**: As discussed in the previous chapter, applications must use encryption for data at rest and secure protocols such as HTTPS and TLS to secure data in transit. Using any custom, deprecated, or weak cryptographic algorithms, protocols, or crypto keys can make the application vulnerable. The certificates and trust chain also need to be periodically checked and validated.

- **Injection**: Applications that do not validate the user-supplied data or inputs before executing the necessary commands are subject to injection flaws, especially when this user-provided data is run as commands or queries against backend databases without sanitization by the application. The attacker's hostile data can trick the application and access data without proper authorization.

- **Insecure Design**: A missing or ineffective control is a bad design that leads to security risks that cannot be protected, even with proper implementation. The primary factor that contributes to this is a lack of understanding of the business requirements related to **confidentiality, integrity, and availability (CIA)** that the application needs to meet and the related profiling of the risks. For a multi-tenant solution, missing controls for tenant isolation of data and functionality can be treated as insecure design.

- **Security Misconfiguration**: This is caused by missing security hardening, control across the application stack, or improperly configured permissions on cloud services. Often, developers leave default insecure configurations or do not harden or change the default IDs or passwords to gain access to critical services or data.

- **Vulnerable and Outdated Components**: Cloud applications use services, components, and libraries from multiple vendors and clouds. Any obsolete library or any component that's left misconfigured can leave the application at a higher risk. If the application is relying on an outdated frontend, backend, or database component or is not upgraded on time, this can expose the enterprise to exploits for longer.

- **Identification and Authentication Failures**: If the application authentication is well implemented, it permits automated attacks such as credential stuffing or brute force, where the attacker has access to a list of usernames and passwords. Using weak, default, or commonly used passwords is not safe. If the application does not handle sessions or authentication tokens correctly, it allows attackers to leverage compromised passwords or session tokens to exploit the weaknesses in the application.

- **Software and Data Integrity Failures**: If applications fail to validate and establish trust across all their components, this can lead to integrity failures related to code and infrastructure that can expose sensitive data to unauthorized components. In a cloud environment, applications are continuously updated. In such situations, upgrades without sufficient integrity verification can provide a backdoor for hackers to distribute and run their installations. This can lead to applications that handle sensitive information being compromised.

- **Security Logging and Monitoring Failures**: Many enterprises find it hard to detect foreigners in the network or that they have been breached. This is mainly because of insufficient logging and monitoring. Many applications fail to log key auditable events such as failed logins or high-value transactions. Similarly, application and API logs are not monitored for abnormal or suspicious activities.

- **Server-Side Request Forgery** (SSRF): These errors occur when the application tries to fetch resources without validating the user-provided URL. This allows an attacker to coerce the application into sending a crafted request to an unexpected destination, even to those behind firewalls or network protection. So, these types of attacks are increasing in the cloud world since this is where the complexity is high and where applications rely on several cloud services.

You can read more about these security risks on the OWASP site, whose URL is provided in the *References* section. SANS Institute is another cooperative for information security thought leadership that specifies the top 25 most dangerous software errors (`https://www.sans.org/top25-software-errors/`). Armed with this information, the security engineer or developer in charge of security can offer insights to the team on conditions, people, or processes that could lead to security incidents.

Depending on the criticality of the system or based on the confidential data handled by the system, enterprises decide to do more complex threat modeling. Complex threat modeling goes beyond just checking for the aforementioned basic security risks. A complex threat model analyzes all the different actors and possible attack vectors. This analysis can leverage the knowledge base of known attack models. MITRE ATT&CK, for example, is a globally accessible knowledge base of adversary tactics and techniques based on real-world observations. The ATT&CK knowledge base is used as a foundation for developing specific threat models and methodologies in the private sector, government, and the cybersecurity product and service community. Enterprises can learn more from the published tactics and techniques since they provide information for many platforms and systems, including information and communication systems, mobile and enterprise operating environments such as Android, iOS, Windows, macOS, and Linux, as well as several IaaS, PaaS, and SaaS services, including containers.

Understanding the security risks and threats earlier in the architecture and design phase can be achieved through security engineering and threat modeling. With these insights, enterprises can relook at the solution for any risk integration, configuration, or processes while the developer can address the code with correct programming. Secure coding guidelines and best practices are shared with the development team to prevent these software errors and attacks. Through automated security testing, many of these errors can be detected and fixed in the earlier days of development. A security bug released to the field is often costlier than fixing it in the development phase.

Known uses

The following are the known uses:

- The *IBM Secure Engineering Framework* discusses the elements of secure product development and a secure supply chain. The framework provides insights into how IBM has implemented secure engineering for its products and services in the core areas of software development. Enterprises can leverage these best practices and implement the patterns for their hybrid cloud application development.

- One of the pillars of AWS's Well-Architected Framework is security. The security pillar focuses on protecting information and systems. Key topics include the confidentiality and integrity of data, managing user permissions, and establishing controls to detect threats and security events.

- The Microsoft Threat Modeling Tool allows software architects to identify and mitigate potential security issues early in the development cycle. Leveraging tools for threat modeling makes it easier for all developers, even those without security backgrounds, to get clear guidance on creating and analyzing threat models.

The DevSecOps pattern

Let's get started!

Problem

How can we incorporate security into the development and operations phases?

Context

Cloud applications are developed, deployed, and managed seamlessly by bringing together development and operations. This enables enterprises to incorporate feedback from application end users as well as push out new features and functionality.

In the earlier development model, security was incorporated at the end of the application development phase. But when application code is continuously updated and deployed, security needs to be incorporated from the early stages of development to ensure it is running on a safe platform and is free from vulnerabilities. The enterprise needs to evaluate the operational risks, while the business owners need to be upraised about the same.

The security team can do security testing to ensure there are no vulnerabilities or weaknesses in the system before promoting the application to production. The checks that are captured in the secure engineering and threat modeling steps need to be run against the application to ensure there are no weaknesses in the application that can be exploited by bad actors to abuse the system or steal data. These checks include looking for the correctness of the design, coding, and integration of software, systems, and services. Some of these checks are static before the application is deployed, while others need to be performed dynamically against the running application. Performing these manually is not possible when multiple versions of the code need to be pushed into production several times a day or week.

Solution

Automating the secure engineering and validation checks is the solution to accelerate security testing. This means that the tools and techniques for validating secure coding and configuration need to be integrated with the CI/CD pipeline.

As shown in the following diagram, when security processes, tools, and capabilities are included in a DevOps pipeline it becomes a DevSecOps pipeline. Some of the security tools that are typically included as part of this pipeline are shown in the following diagram:

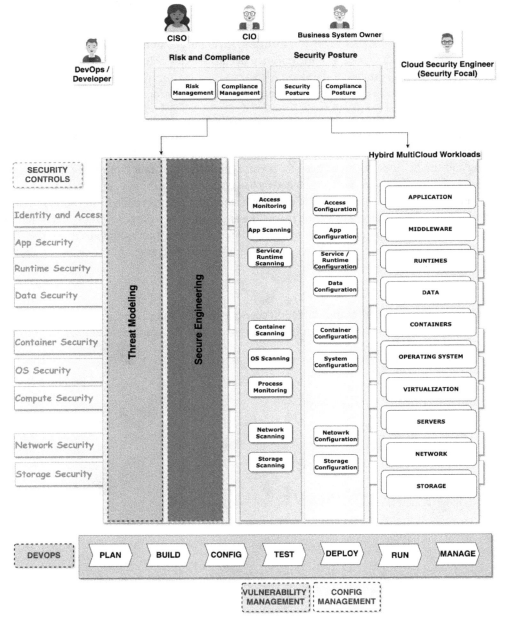

Figure 8.3 – DevSecOps

Source and byte code scanning tools

Application developers need a way to verify that their code and application stack is free from vulnerabilities and threats, such as those shared by OWASP. These tools analyze source code to determine vulnerabilities and poor coding practices. These tools can also trace user input through the application (such as code flow analysis and taint propagation) to uncover various injection-based attacks. If the source code of the component or library is not available, byte code scanners, if available, are used for the analysis. Binary security analyzers analyze the application binary files in place of source code. Any dependencies and risks related to compilation and build environments also surface during this phase.

Dynamic security scanning tools

These tools analyze the application when they're running. They treat the application as a black box and, without knowledge of the internals, try to break into the application.

Runtime analysis and code coverage tools

Runtime analysis tools try to understand the application behavior from a system, operations, and performance perspective. Code coverage tools provide insights into the quantity of code that's tested while the unit tests are executed. By combining the insights from these tools with monitoring resource usage, you can understand the application during its running state, including memory used and disk I/O concerning the application's expected behavior. Code coverage is a generic software quality matrix that tells us how much of the source code is covered by a set of test cases, and it also gives us insights into runtime security analysis. Analyzing the application from the black box and white box perspectives can provide a greater understanding of the potential security issues that may exist, such as load or stress testing and **denial-of-service (DoS)** attacks.

Vulnerability management

Vulnerability management is a practice that is shared between development and operations. When an application moves from development to production, it is important to ensure that it is free from vulnerabilities and is deployed securely on a safe platform. In an agile model for the CD of applications, the new code rollout should include updated functions as well as bug fixes. A major release typically includes multiple changes and features; a minor release mostly has limited enhancements and bug fixes. In these releases, it is important to confirm that there are no security impacts or risks. If any of the changes have a security impact, they need to be reviewed and approved to ensure the operational risk is communicated.

Vulnerability management deals with managing vulnerabilities when the application is used or deployed. This is applicable across infrastructure, operating systems, application runtimes, and services leveraged from the cloud. Scanning these systems during or post-deployment as part of DevSecOps is a recommended practice.

Enterprises subscribe to notifications related to discovered vulnerabilities in the most commonly used cloud and non-cloud infrastructure, software, and service components. Then, they evaluate these notifications to understand if any of their applications or systems are impacted because of the discovered vulnerability. A plan is put in place to fix the vulnerability or security defect promptly. These vulnerabilities may exist in any part of the stack, including the following areas:

- Infrastructure – compute, network, and storage

- Applications and runtimes

- Software components and cloud platform services such as database and container services

How often the vulnerability assessment occurs is established based on the components, while fixing the vulnerability is based on the cloud consumption and delivery patterns. The shared responsibility model determines who owns the vulnerability management duties. Certain cloud platforms provide automated security vulnerability and patch management capabilities that simplify this process for cloud consumers. Typically, specific engineers are assigned security focal roles to confirm the readiness of each component deployed to the cloud. They also need to ensure that the service conforms to security and compliance readiness. If the cloud application is using third-party components, the providers of the software components share the notification regarding any vulnerabilities that have been detected in their components. The security focal for the application needs to action the security fix as per the timelines proposed. Sometimes, this may result in application downtime.

Containers are an integral part of cloud-native application development. As more and more applications are designed in a microservices model, the capabilities are packaged, deployed, and operated as containers. Organizations leverage these capabilities to discover vulnerabilities and compliance policy problems in containers. Again, these capabilities are integrated with the CI/CD pipeline, where the container vulnerability detection service discovers critical vulnerabilities in container images or containerized applications. As soon as a new container image is built and pushed into the container repository, the image gets automatically scanned for vulnerabilities. The scanning process can evaluate the security posture of the image against the organization's policies, as well as against the database of known issues. Based on the results of the vulnerability scan, the deployment process can be allowed or blocked. If several high-severity critical issues are reported, the CI/CD process is instructed to abort the deployment. The scanning process also checks for violations against audit guidelines. The results provide details of vulnerable packages and security violations. The security focal needs to work with the application development team to fix these issues based on their criticality before the application can be deployed in production. Some scanning solutions provide recommendations for how to quickly fix the application security issues. Low or fewer observations are treated as warnings and may be addressed later.

Vulnerability and patch management should cover both in-house components as well as supply chain components. The components of cloud applications are sourced from multiple sources, including open source and third-party software vendors. The security controls for the supply chain of these components are critical to developing and delivering a secure cloud application that's free from vulnerabilities.

Every enterprise should have a model for reviewing and approving the components before they're used in the application. Any vulnerabilities that are reported or found as part of the review or scan need to be recorded and tracked to closure.

Security configuration management

Security configuration management is the process of validating and improving the operational controls in place for a secure and safe deployment. This process needs to be shared between the development and operations teams and needs to be applied across the stack while covering the infrastructure, platform, and application layers. Secure deployment ensures the application deployment, pipeline, and deliverables are configured. Configuration management looks for errors in the configuration and settings across these components. It identifies misconfigurations that can result in a security event, non-compliance, or a vulnerability being exposed that may be exploited by a hacker. Configuration issues or errors can occur in any of the following components or scenarios:

* Default credentials or predefined passwords that have been set for operating systems or application data stores

* Storage is left unencrypted, making it easy for attackers to gain unauthorized access to the enterprise's data

Known uses

The following are the known uses:

* Microsoft DevSecOps controls list the security controls that are recommended to be integrated into each stage of a CI/CD DevOps process.

* AWS provides a model for building an end-to-end AWS DevSecOps CI/CD pipeline with open source SCA, SAST, and DAST tools.

* IBM provides AI-powered automation capabilities for DevOps and a suite of DevSecOps-ready tools and services to enable secure continuous delivery, integrated security testing, and cloud-native delivery pipelines.

* The OWASP project provides a **Software Composition Analysis (SCA)** tool that attempts to detect publicly disclosed vulnerabilities contained within a project's dependencies.

* SonarQube (SAST) provides capabilities for catching bugs and vulnerabilities in the application through automated static code analysis rules.

* IBM QRadar Vulnerability Manager is a network scanning platform that detects vulnerabilities within applications, systems, and devices.

* IBM Security Guardium Vulnerability Assessment scans data infrastructures (databases, data warehouses, and big data environments) to detect vulnerabilities and suggest remedial actions.

- Clair is an open source project that provides a tool for monitoring the security of containers through the static analysis of vulnerabilities in container images.

- Aqua provides products that automate security testing and continuously scan registries and serverless function stores to detect vulnerabilities and risks. Aqua's advanced vulnerability scanning and management DevOps can detect vulnerabilities, embedded secrets, and other risks during the development cycle, as well as prioritizing mitigation via risk-based insights to realize DevSecOps.

- AWS Config is a service that's used to assess, audit, and evaluate the configurations of cloud resources. Config continuously monitors and records your AWS resource configurations and allows you to automate the evaluation of recorded configurations against desired configurations. This enables you to simplify compliance auditing, security analysis, change management, and operational troubleshooting.

- IBM Cloud Security and Compliance Center is a central place for managing security and compliance controls that help define and enforce resource configuration standards. It tries to standardize resource configuration across deployments by defining enforceable rules and templates.

- CloudCheckr is a security solution that can monitor for secure AWS and Microsoft Azure cloud configuration to maintain posture and ensure governance.

Summary

Secure DevOps or DevSecOps integrates security capabilities and tools into a DevOps pipeline and automates the checks to ensure the application's design, code, and components are free from vulnerabilities. In this chapter, we looked at the top security issues that need to be addressed by the application, runtime, and services. DevSecOps, along with the pattern of implementing secure engineering and threat modeling earlier in the application development cycle, helps shift eft security. Vulnerability management requires continuously monitoring security issues related to infrastructure, operating systems, application runtimes, and services. Timely patching of the vulnerabilities is required to keep the application safe from attacks and meet the audit and compliance guidelines. Security threats and technical vulnerabilities continue to evolve. With several services to deal with across clouds while building the application, manually ensuring all the configurations are correct is not possible. A sustainable solution is an automated security configuration management process that can identify misconfiguration issues and fix them. Good security requires a combination of secure engineering and an awareness of security threats and risks through threat modeling, automated vulnerability, and configuration management practices. The automation aspects for vulnerability and configuration management also help address the skills gap for an enterprise that does not have enough security engineers.

In the next chapter, we will learn about protecting data in use.

References

Refer to the following resources for more details about the topics that were covered in this chapter:

- *Security in Development – The IBM Secure Engineering Framework*: https://www.redbooks. ibm.com/redpapers/pdfs/redp4641.pdf

- *AWS Well-Architected*: https://aws.amazon.com/architecture/well- architected/

- *Microsoft Threat Modeling Tool*: https://docs.microsoft.com/en-us/azure/ security/develop/threat-modeling-tool

- *DevSecOps controls*: https://docs.microsoft.com/en-us/azure/cloud- adoption-framework/secure/devsecops-controls

- *OWASP Dependency-Check*: https://owasp.org/www-project-dependency- check/

- SonarQube – *Code Quality and Code Security*: https://www.sonarqube.org/

- AWS provides a model for building an end-to-end AWS DevSecOps CI/CD pipeline with open source SCA, SAST, and DAST tools: https://aws.amazon.com/blogs/devops/ building-end-to-end-aws-devsecops-ci-cd-pipeline-with-open- source-sca-sast-and-dast-tools/

- IBM – *DevSecOps*: https://www.ibm.com/in-en/cloud/learn/devsecops

- *Overview of Qradar Vulnerability Manager*: https://www.ibm.com/docs/en/qradar- on-cloud?topic=manager-overview-qradar-vulnerability

- *IBM Security Guardium Vulnerability Assessment*: https://www.ibm.com/in-en/ products/ibm-guardium-vulnerability-assessment

- *Scanning container image vulnerabilities with Clair*: https://www.redhat.com/en/ blog/scanning-container-image-vulnerabilities-clair

- Aqua – *Vulnerability Scanning and Management*: https://www.aquasec.com/products/ container-vulnerability-scanning/

- *AWS Config*: https://aws.amazon.com/config/

- *IBM Cloud Security and Compliance Center*: https://www.ibm.com/cloud/security- and-compliance-center

- *CloudCheckr*: https://cloudcheckr.com/solutions/cloud-security/

Part 5:
Cloud Security Posture Management and Zero Trust Architecture

Cloud Security Posture Management (CSPM) helps to proactively monitor, track, and respond to security incidents to meet the enterprise's **Governance Risk and Compliance (GRC)** requirements. This part will provide patterns for building end-to-end visibility and integration of security processes to get a unified security posture across hybrid cloud environments. This part discusses the principles of a zero trust model and the key components needed to build a reference pattern supporting the model. Further, this section will discuss how to leverage hybrid cloud security patterns for protecting critical applications and data using zero trust security practices.

This part comprises the following chapters:

- *Chapter 9, Managing Security Posture for your Cloud Deployments*
- *Chapter 10, Implementing Zero Trust Architecture with Hybrid Cloud Security Patterns*

Managing the Security Posture for Your Cloud Deployments

An enterprise's **Governance, Risk, and Compliance (GRC)** policies provide guidelines on what the IT systems need to achieve to meet the security and compliance goals. **Cloud Security Posture Management (CSPM)** helps to proactively monitor, track, and react to security violations to meet the GRC requirements. This chapter provides patterns on how to build end-to-end visibility and integration of security processes and tooling throughout the organization to get the security posture for the cloud applications. The security and compliance posture provides a method to understand the security controls implemented and their effectiveness. This chapter discusses how to prepare the enterprise to respond to large volumes of alerts and events related to cloud security. Given the use of multiple tools and shortage of staff, enterprises need to adopt security orchestration, automation, and response to improve their effectiveness against security events. The following diagram illustrates the pattern of CSPM:

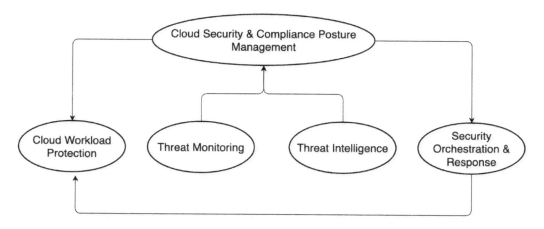

Figure 9.1 – CSPM patterns

The topics covered under this pattern include the following:

- Security and compliance posture management
- **Cloud Workload Protection (CWP) platform**
- Threat monitoring
- **Security Orchestration and Response (SOAR)**

CSPM patterns

Let's get started!

Problem

How to evaluate, design, build, and implement solutions that provide the security and compliance posture necessary to reduce and manage risks to acceptable levels.

Context

The advantages of multicloud are better realized only if the enterprise can take the required risks. The multicloud strategy needs to be complemented with the careful management of threats and operational efficiencies. Managing multiple clouds brings in more challenges, such as ensuring value for money, as well as meeting the audit, compliance, and regulatory requirements. These risks will cut across all layers, including governance, application, integration, data, and even insider threats.

Multicloud governance provides the guardrails or set of controls to improve efficiency and eliminate risks. The controls that IT systems should implement to meet the security and compliance goals are defined in the corporate security policy document. This policy document is typically generic in nature and mostly discusses only security controls design and implementation for one's own data centers. As enterprises adopt more clouds, specific controls for each cloud are incorporated. As discussed in *Chapter 2*, some of these controls need to be met by the cloud provider, while others are to be met by the enterprise as the cloud consumer. These security policies also need to be specified for different consumption and delivery models, namely **Infrastructure as a Service (IaaS)**, **Platform as a Service (PaaS)**, and **Software as a Service (SaaS)**. The security controls are defined for each layer or security subdomain, namely the following:

- Identity and access management
- Infrastructure – storage, compute, and network
- Applications
- Data
- Platforms and runtimes

- DevSecOps
- Personal security
- Physical security

Multicloud compliance is about meeting the requirements or criteria that are needed by the industry, country, or domain that the enterprise is operating in. There is a variety of compliance requirements across each layer listed previously based on the consumption model. These controls are needed not only to meet the compliance requirements but also to pass the audit by certification agencies.

Enterprises need a way to visualize the security and compliance controls and their effectiveness in one place. Also, in multicloud deployments, the scale is bigger in terms of the millions of events and incidents that need to be analyzed in near-real time and reported to the leadership in a timely manner. Any ineffective controls, or failure of security controls, not addressed in time can result in the whole business being impacted.

To centrally provide these insights, the solution needs to collect and analyze events and logs from across all the cloud and non-cloud environments and across all layers or subdomains – identity and access, infrastructure, application, and data. This will often be quite challenging from an integration perspective and overwhelming in terms of the volume of security data. Most cloud service providers do not have a standard way of sharing security data in a dashboard or with a customer tool for further analysis. The same challenge is true for third-party security tools and products. Another challenge is how the cloud policies are administered centrally and enforced uniformly across the cloud and non-cloud environments. From a security operations and management perspective, finding employees with skills and experience in working with different cloud and various security tools is another challenge.

Solution

Enterprises use CSPM as the solution to understanding the key security and compliance risks in the cloud. Given the dynamic nature of the cloud, the CSPM solution needs to have a mechanism to continuously monitor the controls. The controls are specified in the security policy document. CSPM plays a critical role in detecting security issues early and securing the multicloud environment by reducing the possibilities of data breaches. We discussed in the previous chapter how and why vulnerabilities and misconfigurations are major entry points for attacks.

As shown in *Figure 9.2*, the CSPM pattern brings threat monitoring, CWP, and security response together. The main objectives of this pattern are as follows:

- An effective way for the enterprise to centrally detect, manage, and act on security risks across multicloud and non-cloud environments
- Early detection of vulnerabilities and misconfigurations, as well as policy violations that can lead to security incidents

- Evaluate and compare the cloud settings against best practices and take action to set it right
- Do a mapping of the controls against the industry, regulatory, and audit controls framework to provide the real-time security and compliance posture of the deployment

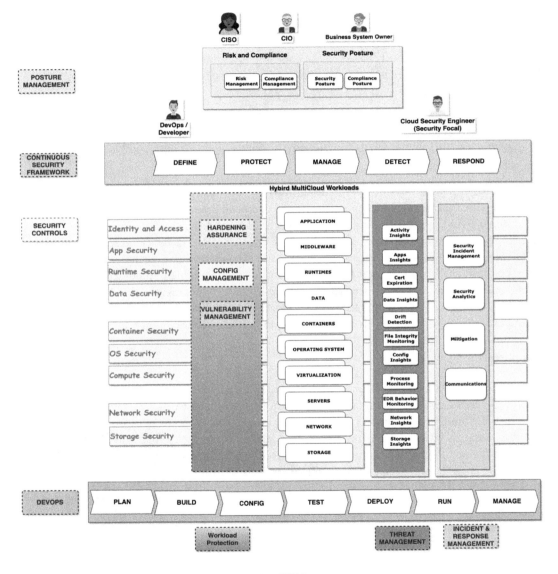

Figure 9.2 – CSPM patterns

The core component of CSPM is a dashboard that provides the security posture on how enterprise workloads are performing against the defined benchmarks, policies, and best practices. The dashboard provides the visibility to discover the risks and continuously monitor the threats to the cloud resources. This dashboard for a hybrid multicloud CSPM solution needs to bring together a security and compliance posture for all the workloads deployed across multiple clouds and non-cloud infrastructures. The unified dashboard or console also provides for centralized security management apart from visibility.

For integration, there is a need to follow a standard schema and method for reporting the security event. The solution needs to be backed by a robust analysis component that will cut out the noise and surface only the important actionable insights onto the dashboard. This way, CSPM also provides a way to do proactive vulnerability, configuration, and threat management. The dashboard can be further personalized for specific roles, such as for a security-focused role that may be responsible for a specific application or set of applications. This kind of dashboard will help those focused on security, even without specific knowledge of individual security tools, respond to the event faster and effectively.

As shown in *Figure 9.3*, for a multicloud CSPM solution, enterprises will need to integrate insights from cloud vendor-specific CSPM solutions. An effective implementation of this CSPM pattern requires the integration of insights from not only cloud providers but also all the third-party threat monitoring and security tools used by the enterprise. This will provide a complete visualization of risks across multiple clouds:

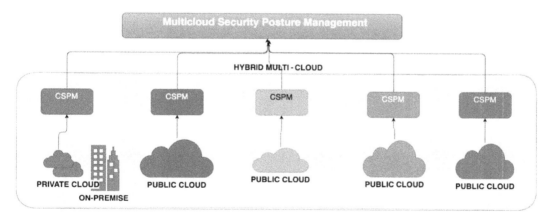

Figure 9.3 – Multicloud security posture management

For continuous security and compliance assurance, and the monitoring of the threats and the effectiveness of the controls, the CSPM needs to be supported by an effective threat monitoring solution. The threat monitoring solution is the engine that powers the dashboard with the right data. It comprises components that need to go hand in hand with the CSPM solution to continuously monitor the threats against the resources and domains. The events across each layer and domain of the cloud stack that we discussed in the previous chapters need to be brought together and analyzed for threats. A typical cloud security stack includes components as discussed in the NIST Cybersecurity Framework. See the link provided in the *References* section for more details. The stack includes, but is not limited to, the following items:

- **Identity and access management**: Threats and events related to the authentication and authorization of IAM activities, such as the overprovisioning of access to identities, user anomalies, and failed logins.

- **Infrastructure (compute, storage, network)**: Vulnerabilities in containers or packages, hypervisor- and virtual machine-related vulnerabilities and risks, unsecured or unencrypted storage buckets, network traffic anomalies, and behavior that includes inbound or outbound traffic to suspicious IPs. The infrastructure will also involve endpoints. **Endpoint Detection and Response (EDR)** is an integrated endpoint security model that combines real-time continuous monitoring and collection of endpoint data with rules-based automated response and analysis capabilities.

- **Data security**: Events related to encryption and key management activities, database activity monitoring, certificate expiry, and so on.

- **Application security and Dev-Sec-Ops**: Web application firewall events and automated application static and dynamic security scan results.

- Events from other **Security Information and Event Management (SIEM)** and SIEM tools.

- Physical security events, including people security, hardware, data center facility-related events, such as a loss of power, air conditioning, and so on, as well as reporting on any events related to the integrated supply chain.

- Also, for the compliance section of the CSPM dashboard, the threat monitoring component needs to continuously verify and confirm that the right security controls are in place and whether they are effective. Any deviations need to be reported.

Based on the identified security gaps, the CWP addresses the security and protection at each layer of the stack. This protection when delivered out of a platform is referred to by IT research and analysis firms such as Gartner as a **Cloud Workload Protection Platform (CWPP)**.

CSPM also works with **Cloud Access Security Brokers** (**CASBs**), which sit in the middle of on-premises and cloud environments and enforce security policies. There are two use cases that we need to deal with from an enterprise usage of multicloud perspective. One is called the **enterprise outbound** model, where the applications of the enterprise connect to cloud services outside the enterprise. The other use case is an enterprise inbound scenario, where the cloud or external applications connect to a service running inside the enterprise. The CASB model mainly looks at enterprise outbound scenarios where enterprise users access cloud resources. The CASB works as a cloud-based security policy enforcement point placed between the cloud service consumers and cloud service providers. The cloud defender model of CASBs also covers enforcing security for cloud inbound scenarios including authentication (SSO), authorization, context-based access, as well as alerting on anomalies and malware detection.

Finally, based on the detection of the incidents, there needs to be an automated way to respond to the security event. The steps taken to respond to the events are captured as part of the security incident response. In some cases, these incident response steps are orchestrated or automated to take rapid and immediate action. The method to alert on critical incidents and orchestrate an automated response is an extension of the CSPM solution and is called SOAR.

SOAR automates the manual task of responding to high-priority security events. Typically, as shown in the following figure, the workflow for remediation of the security incident is typically initiated through integration hooks or alerting mechanisms. The integration hooks can be anything such as creating a ticket in task management tools such as Jira for the security team or creating an alert in notification systems such as PagerDuty or Slack. On receiving the alert, SOAR implements dynamic playbooks or scripts that are executed to resolve the incident. The response may be adapted for purpose depending on the incident. SOAR platforms also include the analysis capability to visualize and understand relationships across incidents. For incidents that are not resolved through automated response, a case may be opened in a case management system such as ServiceNow for further investigation. Security case management provides a means for security analysts who are engaged in threat hunting to gather information on suspicious activity in their environment. Case-related records, such as security incidents, observables, CIs, and affected users, can be added to cases to accommodate broad and specific analysis.

As shown in *Figure 9.4*, this pattern includes the components for threat monitoring and security intelligence, investigation, case management, and SOAR working together to detect threats early and fix them as soon as they are detected, reducing the blast radius. The pattern is powered by a data lake that allows the ingestion of security data from various sources. Security experts can dip into this data lake and make queries or federated searches to narrow down or correlate events across their hybrid multicloud environment:

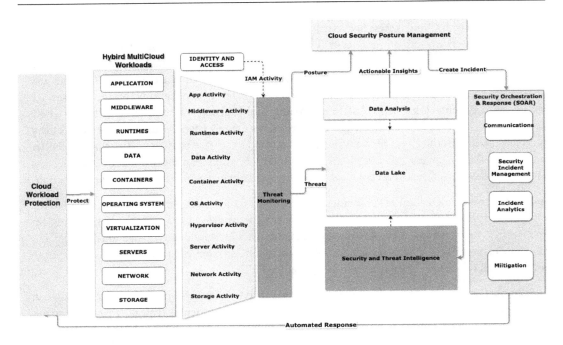

Figure 9.4 – CSPM protect-detect-respond loop

Threat monitoring systems are often backed by a threat intelligence component. We will find several billion events generated by cloud resources on a daily basis. Deriving actionable insights from these several billion events manually is not easy. Most CSPM solutions are backed by security intelligence systems that use machine learning, behavioral analytics, and application-based intelligence to detect threats faster. These insights and intelligence are often shared, compared, looked up, or aggregated in a threat intelligence platform. Threat intelligence provides observations on cloud resources and artifacts, such as IP addresses, domains, and file hashes with known cyber threats. The threat intelligence platform provides an API that enables customers to programmatically access threat and security information. This intelligence is combined with the security investigation to get actionable insights. These platforms provide standards-based data connectors to exchange threat indicators and threat intelligence. The popular open source standard in the industry for the exchange of threat information is STIX/TAXII. **Trusted Automated Exchange of Intelligence Information (TAXII™)** is an application protocol for exchanging **Cyber Threat Intelligence (CTI)** over a secure protocol. **Structured Threat Information Expression (STIX™)** is a language and serialization format used to exchange CTI.

Known uses

Here are some of the known uses:

- AWS provides a security hub that helps automate AWS security checks and centralize security alerts. The CSPM solution provides automated checks based on a collection of security controls and simplifies compliance management with built-in mapping to different frameworks, such as CIS and PCI DSS. AWS uses **AWS Security Finding Format** (**ASFF**) to ingest and correlate ingested findings across products to prioritize the most important ones. The security hub consumes, aggregates, organizes, and detects deviations and prioritizes findings using this normalized format. The security hub includes ways to enrich these findings, remediate them, or provide information about them to ticketing and task management systems.

- Microsoft Defender for Cloud is a CSPM and CWPP solution for all of the Azure, on-premises, and multicloud (AWS and GCP) resources. Defender for Cloud also provides a secure score for continual assessment of the security posture. This helps with tracking new security opportunities, as well as reporting on the progress of improving the posture. The score is based on the **Azure Security Benchmark** (**ASB**), which provides best practices and recommendations to help improve the security posture of workloads, data, and services. The regulatory compliance dashboard provides insights into the compliance posture based on how the deployment meets specific compliance requirements. Several regulatory standards, common frameworks, and industry guidelines are compared, such as PCI-DSS, SOC TSP, NIST, HIPAA/HITRUST, ISO 27001:2013, and FedRAMP. The CWPP part recommends actions to protect the workloads from known security risks, as well as defending in real time to prevent security events from developing.

- Security Command Center is the security and risk management platform for Google Cloud that provides a single, comprehensive dashboard to get centralized visibility and control related to misconfigurations and vulnerabilities, compliance, and threats related to Google Cloud assets.

- IBM Cloud Security and Compliance Center provides a unified experience to view, manage, and govern cloud resource configurations and centrally manage compliance to enterprise and regulatory guidelines. Integrations are possible between IBM Cloud Security and Compliance Center and various services. Enterprises can integrate third-party tools such as Tanium™ Comply to view all your compliance data in one location or integrate with the **OpenShift Compliance Operator**, also called **OSCO**, to scan Red Hat clusters against defined OSCO profiles. Like other CSPM solutions, the objective is to define configuration rules and templates that prevent the unsecure configuration of cloud resources. The solution also allows the investigation of the results and can be shared with auditors.

- CSPM solutions from vendors such as Zscaler go beyond providing visibility to automated remediation – such as fixing the misconfiguration or shutting down a vulnerable container. Zscaler provides a dashboard to report the security posture across AWS, Azure, and Google Cloud.

- Orca Security provides a CSPM tool that works on AWS, Azure, and Google Cloud services. Orca Security combines CSPM and `CWPP` capabilities. The goal is to provide visibility and analysis in a multicloud environment.

- Trend Micro Cloud One Conformity is a CSPM tool that works across AWS and Azure environments to provide security governance and compliance capabilities.

- The Colortokens Xcloud solution provides a solution that provides complete visibility of multicloud environments and automates the inspection of all layers of the cloud stack for common configuration mistakes and prioritizes them based on risk. The solution continuously monitors configuration changes to identify security risks.

- The Aqua CSPM solution can scan, monitor, and remediate configuration issues in public cloud accounts according to best practices and compliance standards, across AWS, Azure, Google Cloud, and Oracle Cloud.

- Prisma Cloud by Palo Alto Networks is a comprehensive **Cloud-Native Security Platform (CNSP)** that provides security and compliance coverage for infrastructure, applications, data, and all cloud-native technology stacks throughout the development life cycle. Prisma Cloud supports cloud operations across hybrid and multicloud environments, all from a single, unified solution, using a combination of cloud service provider APIs and a unified agent framework.

- As discussed in the *Solution* section, threat monitoring is often backed by a threat intelligence component. Some of the popular implementations of the threat intelligence components are as follows:

  - IBM X-Force Exchange is a threat intelligence sharing platform that you can use to research security threats, aggregate intelligence, and collaborate with peers.

  - Amazon GuardDuty is a threat detection service that continuously monitors workloads on the AWS cloud for malicious activity and delivers detailed security findings for visibility and remediation.

  - Microsoft Sentinel is a threat intelligence platform that can be applied to security products and used to detect potential threats to an enterprise. Microsoft Sentinel provides threat indicators to help detect malicious activity observed in the cloud environment and provide context to security investigators to help inform response decisions.

  - **Google Cloud Threat Intelligence (GCTI)** for Chronicle is a service that creates actionable insights on threats. The Chronicle GCTI API enables customers to programmatically access their security data directly through API calls to the Chronicle platform and use this pool of curated, high-fidelity threat information for the investigation of their security cases.

Summary

In this chapter, we learned about how to use the CSPM pattern to create a security and compliance dashboard. The dashboard is powered by threat monitoring and intelligence components that need to continuously ingest and monitor events and activities across all clouds and on-premises resources. Threat intelligence can be used to narrow down the huge volume of data to create actionable insights. Responding to events in time is equally important as detecting threats. Security automation and incident response patterns help provide an automated response to threats and vulnerabilities and accelerate incident response.

In the next chapter, we will learn how to put together all the patterns that we have discussed so far to build security solutions in a zero-trust model.

References

Refer to the following for more information about the topics covered in this chapter:

- NIST Cybersecurity Framework `https://www.nist.gov/cyberframework/getting-started`

- *Market Guide for Cloud Workload Protection Platforms*: `https://www.gartner.com/en/documents/4003465`

- *Enterprise Cloud Forensics and Incident Response* | SANS FOR509: `https://www.sans.org/cyber-security-courses/enterprise-cloud-forensics-incident-response/`

- Cloud security posture management – AWS Security Hub – Amazon Web Services: `https://aws.amazon.com/security-hub/`

- **AWS Security Finding Format** (**ASFF**): `https://docs.aws.amazon.com/securityhub/latest/userguide/securityhub-findings-format.html`

- Microsoft Defender for Cloud: `https://docs.microsoft.com/en-in/azure/defender-for-cloud/defender-for-cloud-introduction`

- Overview of the Azure Security Benchmark: `https://docs.microsoft.com/en-us/security/benchmark/azure/overview`

- Security Command Center: `https://cloud.google.com/security-command-center`

- IBM Cloud Security and Compliance Center: `https://www.ibm.com/cloud/security-and-compliance-center`

- Zscaler CSPM for workloads: `https://www.zscaler.com/resources/zscaler-for-workloads`

- Zscaler posture control for cloud-native applications: `https://www.zscaler.com/products/posture-control`
- Colortokens CSPM: `https://colortokens.com/about-cspm-cloud-security`
- Aqua Security CSPM: `https://www.aquasec.com/products/cspm/`
- Prisma Cloud by Palo Alto Networks: `https://www.peerspot.com/products/prisma-cloud-by-palo-alto-networks-reviews`
- Threat detection with AWS Cloud: `https://aws.amazon.com/security/continuous-monitoring-threat-detection/`
- Introduction to TAXII: `https://oasis-open.github.io/cti-documentation/taxii/intro.html`
- Introduction to STIX: `https://oasis-open.github.io/cti-documentation/stix/intro`
- IBM X-Force Exchange threat intelligence: `https://exchange.xforce.ibmcloud.com/`
- Understand threat intelligence in Microsoft Sentinel: `https://docs.microsoft.com/en-us/azure/sentinel/understand-threat-intelligence`
- Intelligent threat detection with Amazon GuardDuty: `https://aws.amazon.com/guardduty/`
- IBM Cloud Pak for Security integrated security platform: `https://www.ibm.com/in-en/products/cloud-pak-for-security`
- IBM Security QRadar SOAR platform: `https://www.ibm.com/in-en/qradar/security-qradar-soar`
- ServiceNow Security Case Management: `https://docs.servicenow.com/bundle/quebec-security-management/page/product/threat-intelligence-case-management/concept/case-mgmt.html`

10

Building Zero Trust Architecture with Hybrid Cloud Security Patterns

This chapter discusses the reference architectures and patterns for implementing the zero trust model. The principles for zero trust are also discussed in detail. We will discuss use cases requiring the zero trust model and how to use the hybrid cloud security patterns to protect critical data using zero trust security practices. The following diagram illustrates the zero trust pattern:

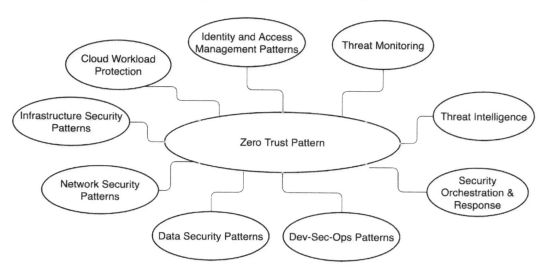

Figure 10.1 – Zero trust pattern

The topics discussed in this chapter include the following:

- Zero trust model
- Principles of zero trust
- Reference architecture for zero trust
- How to use hybrid cloud security patterns to take a zero trust approach

Zero trust pattern

Let's get started!

Problem

How to implement security for a dynamic, distributed, and varied infrastructure, users, endpoints, applications, and data deployed in a hybrid multicloud environment.

Context

With digital transformation, the way enterprises operate has changed. The enterprise boundaries have become thin, enabling business collaboration across customers and partners. Most enterprises use multicloud as a strategy for innovation at speed, scale, performance, availability, and resiliency.

As shown in the following diagram, a typical enterprise IT landscape consists of many resources spread across multiple clouds accessed by users using multiple channels, such as web or mobile, using various endpoint devices. Enterprise applications and data are also shared with their customers and partners:

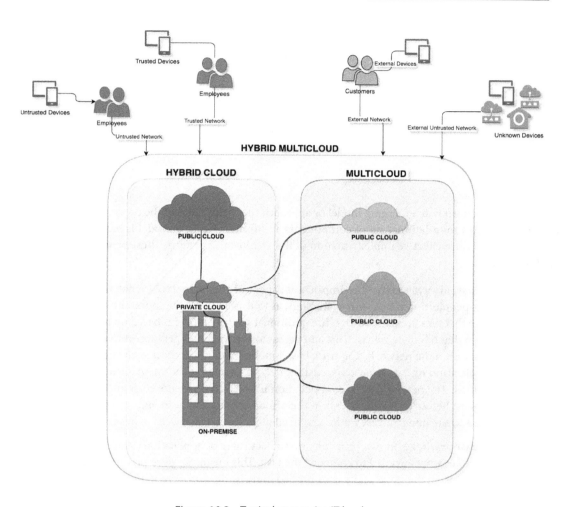

Figure 10.2 – Typical enterprise IT landscape

Traditional security that was more focused on perimeter protection has become obsolete or deficient in its capabilities to protect diverse and distributed assets. This model of perimeter-based network security controls, such as firewalls and VPNs, automatically trusts users who are inside the network. But with the vanishing perimeters and collaboration across enterprises the threats can arise from within the internal network. Also, post-pandemic, there are more users connecting to the enterprise application remotely from their homes or unknown networks. Enterprises need to enable secure access to enterprise resources for remote users and workers. Enterprises now also need to be able to rapidly form (and dissolve) partnerships. At the same time, they need to protect the internal network from insider attacks, as well as managing the risk of resource access from unmanaged and jailbroken devices. Perimeter-based defense is not equipped to address this use case. **Insider threats** are another macro trend or use case that enterprises are trying to address. Enterprises, especially in the blockchain, cryptocurrency, defense, and financial service industries, need protection or look to solve this issue as part of their modernization. All these lead to high chances that the traditional approach of centralized protection mechanisms will be breached. The biggest risk with the traditional model is that it will not be able to sustain the dynamic changes in terms of variety and scale in a hybrid multicloud environment.

Innovating at speed requires rethinking security and compliance in the multicloud context. There is a need to look at the security model differently to mitigate risks with assurance and demonstrate continuous compliance and protection against threats.

Solution

The zero trust pattern is an emerging model or approach that helps address the previously discussed challenges using a foundational set of components and functions. This model is based on certain principles that ensure effective implementation of the architecture pattern. These principles include the following:

- **Never trust, always verify**: The important aspect of this principle is that you don't trust anything–people, processess, and technology, as well as network, compute, and storage – until they prove that they are trustworthy. The traditional security model is based on trust. A user or application that has once gained trust and access to the resources gets access to multiple other resources within the network. The trust-based model does not protect against insider attacks on an application or data. In a zero trust model, no user, application, or resource is trusted until verified. The network, user, or application is asked to prove the trust and is verified for every request before providing access to the resource. With this principle, the trusted zone is minimized where implicit trust works, and in all other cases, trust needs to be verified explicitly.

- **Enable least privilege**: In this principle, no subject (user or application) shall be given more privileges than what's needed to complete the task. This means no over-provisioning of access to any resource on the cloud and it is controlled more by the function of the subject as opposed to its identity.

- **Assume breach**: This principle assumes that the trust is fundamentally flawed and all parts of the system are vulnerable to attacks. With an assume breach mentality, the enterprise assumes cyber-attacks will happen and plans the defense strategies accordingly. In this active defense model, the enterprises shift their focus toward continuous monitoring, security, and compliance.

The NIST publication (link provided in the *References* section) discusses the various aspects of the zero trust model focused on protecting resources (assets, services, workflows, network accounts, and so on) and addresses the requirements of accessing any cloud resource from anywhere on any device to provide a reliable and secure experience.

Based on the use cases and general deployment models, the core set of capabilities required to implement a zero trust security pattern includes the following:

- A control plane that provides the way to define and manage security policies. This is often backed by a policy engine and administration capabilities. This is also referred to as the **Policy Administration Point (PAP)** and **Policy Decision Point (PDP)**.

- A system that supports adaptive identity and policy-driven access control. In a zero trust model, the user is not automatically trusted and there is a need to continuously verify and authenticate users no matter where, when, and how they access a system. This identity and access management system identifies the user or application requesting access to the resource and manages the authentication and authorization. The system detects any anomalies in the access patterns and steps up the authentication as needed through **Multifactor Authentication (MFA)**. The system also provides finer authorization controls at the resource level and keeps an audit trail of the overall process. A data plane that includes a mechanism to apply policy-enforced context-specific security for the application and data acts as the **Policy Enforcement Point (PEP)**.

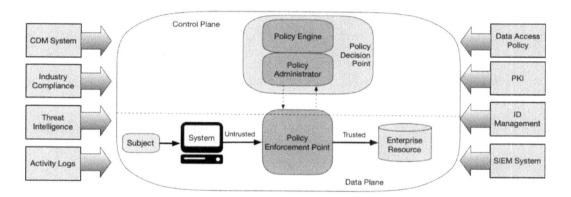

Figure 10.3 – Zero trust pattern (taken from https://nvlpubs.nist.gov/
nistpubs/SpecialPublications/NIST.SP.800-207.pdf)

- Context-specific and policy-enforced application and data security. The context includes several factors, such as the user, device, network, time of day, resource for which access is requested, and classification of the resource based on the sensitivity. Based on the determined context and dynamically updated policies, access is allowed or denied to the resource with technical assurance based on zero trust.

- A zero trust pattern also supports the continuous security model, where the system learns continuously through the observability inputs (logs, monitoring, and tracing) from across all layers from multiple clouds. The analytics of this massive amount of data is often backed by machine learning models that enable the system to decide dynamically on the requests for accessing a cloud resource. The system also automatically updates or adjusts the policies based on the new context and resources identified.

Using hybrid cloud security patterns for zero trust

All the security components in the security architecture need to be rethought – identity, network, app, data, devices, and analytics for the cloud-based workload to implement a zero trust model.

But you can use all the hybrid cloud security patterns that we have learned about in this book to implement a zero trust pattern.

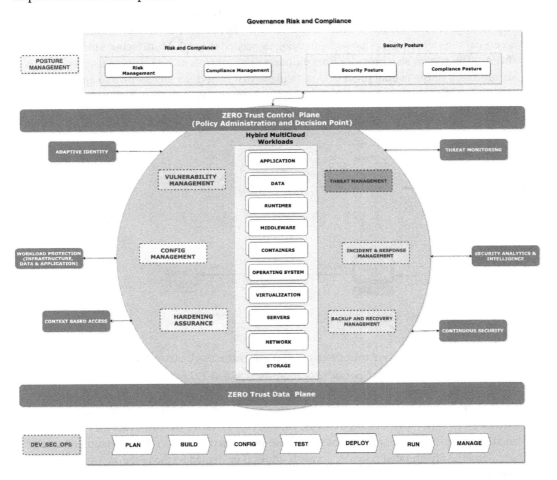

Figure 10.4 – Using hybrid cloud security patterns for the zero trust pattern

The preceding diagram provides the mapping of the several patterns and shows how to overlay them on a zero trust model. The following points discuss the content from previous chapters of this book and how they can be used in a zero trust pattern:

- **Digital transformation and shared responsibility model**: In *Chapters 1* and *2*, we discussed the digital transformation of enterprises and the vanishing of perimeter-based access controls, enabling a collaborative environment across customers, partners, and cloud providers that underlines the need for a shared responsibility model for a zero trust pattern. These chapters also discussed the governance, risk, and policies that hybrid multicloud deployments need to address from an industry and regulation perspective.

- **Identity and access management**: In *Chapters 3* and *4*, we learned about the patterns for cloud and application identity and access management. These chapters covered the standards-based model to address the authentication and authorization needs of a cloud application. These patterns, combined with patterns for the detection of anomalous user behavior based on the risk context across assets – privileged users, devices, data, and so on – can be used as an adaptive identity capability. The idea is to enable least privilege access by discovering and assessing risk across data, identities, endpoints, apps, and infrastructure.

- **Infrastructure security**: In *Chapters 5* and *6*, we learned about patterns to provide isolation to varying degrees and security for bare metal, VMs, containers, and serverless compute types. We also discussed patterns for trusted compute and securing the hybrid cloud network infrastructure. We learned about how to isolate network traffic based on purpose and target environments and network protection for cloud workloads. This provides the core capability in the zero-trust pattern to do microsegmentation and create smaller secure zones.

- In *Chapter 7*, we learned about the core patterns that can be used to protect data at rest, in transit, and in use. These patterns combined with multicloud key orchestrator and encryption can be leveraged to provide context-based access and protection for data. The protection patterns need to follow the never trust, always verify policy with context-aware access control to all apps, data, APIs, endpoints, and hybrid cloud resources

- The zero trust model underlines the need for security to be dealt with early in the development and deployment life cycle, as discussed in *Chapter 8*. Shifting left security with DevSecOps and continuous monitoring of threat and vulnerability management are core tenets of a zero trust pattern. Zero trust mandates the need for continuous security through identifying vulnerabilities and control configurations for cloud resources to facilitate the management of risks.

- Automated response to anomalous behavior and continuous improvement of policies based on threat patterns is a core component of zero trust. Assume breach, identify threats, and automate responses with CSPM and SOAR capabilities, discussed in *Chapter 9*, will not only stop the immediate attack but also dynamically adapt access controls.

- The zero trust model brings together siloed detection of threats and management to a central platform with intelligence to dynamically protect all cloud resources based on context-based policies.

Known uses

The following are the known uses:

- The AWS guiding principles for building zero trust focus on bringing together identity and network capabilities where possible. The model recommends taking a use case-centric approach. There are several use cases, such as workforce mobility, software-to-software communications, and digital transformation projects, that can benefit from the enhanced security provided by zero trust.

- Microsoft integration guidance leverages software and technology partners capabilities to build zero trust solutions. The implementation of the model and principles is discussed in the link provided in the *References* section. Importantly, as shown in *Figure 10.5*, security policy enforcement is at the center of the zero trust architecture. This includes MFA with conditional access, which takes into account user account risk, device status, and other criteria and policies that you set. Policies that are configured for each of these components are coordinated with your overall zero trust strategy. Device policies determine the criteria for healthy devices and conditional access policies require healthy devices for access to specific apps and data. Threat protection and intelligence monitor the environment, surface current risks, and take automated action to remediate attacks:

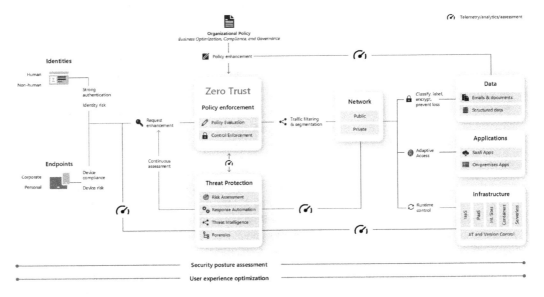

Figure 10.5 – Microsoft zero trust guide (taken from https://docs.microsoft.com/en-us/azure/security/fundamentals/zero-trust)

- Google's BeyondCorp model and a few start-ups have built systems and platforms on top of the zero trust principle to establish the integrity and trust level explicitly. This is based on the organization's risk threshold and tolerance to provide access to assets and data systems. This is a new approach where the traditional perimeter-based protection with firewalls is changed to context-based access. This supports one of the other macro trends that companies are trying to address – secure and remote access for employees to enterprise resources. This requirement has been amplified during the COVID-19 pandemic, when most employees access corporate resources from different places.

- IBM's zero trust framework, as described in the following figure, combines security architecture with a zero trust governance model. The security controls are established and managed at the enterprise level, which combines people, processes, and technology. The framework has the three steps of defining context, verifying and enforcing, and resolving incidents. The **Define Context** step is where the security policies are set. The **Verify & Enforce** step implements policy enforcement. **Resolve Incidents** deals with operations and implements continuous security monitoring solutions and automated responses:

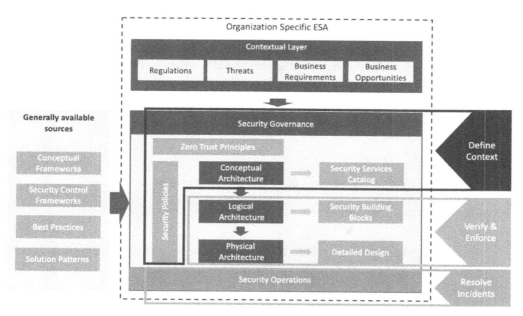

Figure 10.6 – IBM zero trust governance model (taken from https://www.ibm.com/in-en/topics/zero-trust)

- The Open Group has published the *Zero Trust Core Principles*, which introduces zero trust to leaders in business, security, and IT. This whitepaper provides a foundation on the drivers for zero trust, their implications, and the role of zero trust. The Open Group *Zero Trust Commandments* is the document built on the *Zero Trust Core Principles* to present a non-negotiable list of criteria for zero trust. The Open Group guide *Axioms for the Practice of Security Architecture* discusses axioms distilled from years of experience in the practice of security architecture during the formative decades of the discipline. They are designed to be timeless statements with broad applicability, including zero trust, regardless of how digital technologies and threats continue to evolve. The ZTA Working Group of The Open Group is an industry-wide initiative to establish standards and best practices for zero trust as the overarching information security approach for the digital age.

Summary

In this chapter, we learned about the principles of a zero trust model and the key components needed to build a reference pattern supporting the model. We also learned how to use the hybrid cloud security patterns that we discussed in earlier chapters to establish the right level of governance, configurations, controls, policies, protection, and automation and use them to build a solution with the zero trust approach. Hybrid cloud security patterns combined with zero trust models help combat cyber security threats and attacks to accelerate the digital transformation of enterprises. There are more macro and micro patterns that will be discovered in this journey and I shall look to capture them in the subsequent editions and present them to you. I hope you enjoyed reading this book as much as I did writing it. Please get in touch with me with your comments and input.

References

Refer to the following for more information about the topics covered in this chapter:

- NIST publication – *Zero Trust Architecture*: `https://www.nist.gov/publications/zero-trust-architecture`
- Guiding principles for building Zero Trust on AWS: `https://aws.amazon.com/security/zero-trust/`
- *How Zero Trust Will Change Your Security Design Approach*: `https://securityintelligence.com/posts/how-zero-trust-will-change-your-security-design-approach/`
- IBM Zero trust security solutions: `https://www.ibm.com/in-en/security/zero-trust`
- Microsoft Zero Trust architecture: `https://docs.microsoft.com/en-us/azure/security/fundamentals/zero-trust`

- Microsoft Zero Trust Guidance Center: `https://docs.microsoft.com/en-us/security/zero-trust/`

- *Zero Trust and BeyondCorp Google Cloud*: `https://cloud.google.com/blog/topics/developers-practitioners/zero-trust-and-beyondcorp-google-cloud`

- The Open Group ZTA Working Group: `https://www.opengroup.org/forum/security-forum-0/zerotrustsecurityarchitecture`

- The Open Group *Zero Trust Core Principles*: `https://publications.opengroup.org/security-library/w210`

- The Open Group *Axioms for the Practice of Security Architecture*: `https://publications.opengroup.org/g192`

- The Open Group *Zero Trust Commandments*: `https://publications.opengroup.org/g21f`

- *What Is Zero Trust and Why Is it So Important?*: `https://www.cyberark.com/resources/blog/what-is-zero-trust-and-why-is-it-so-important`

- *The Key Components and Functions in a Zero Trust Architecture*: `https://cpl.thalesgroup.com/blog/encryption/key-components-function-in-zero-trust-architecture`

Index

Packt.com

Subscribe to our online digital library for full access to over 7,000 books and videos, as well as industry leading tools to help you plan your personal development and advance your career. For more information, please visit our website.

Why subscribe?

- Spend less time learning and more time coding with practical eBooks and Videos from over 4,000 industry professionals

- Improve your learning with Skill Plans built especially for you

- Get a free eBook or video every month

- Fully searchable for easy access to vital information

- Copy and paste, print, and bookmark content

Did you know that Packt offers eBook versions of every book published, with PDF and ePub files available? You can upgrade to the eBook version at packt.com and as a print book customer, you are entitled to a discount on the eBook copy. Get in touch with us at customercare@packtpub.com for more details.

At www.packt.com, you can also read a collection of free technical articles, sign up for a range of free newsletters, and receive exclusive discounts and offers on Packt books and eBooks.

Other Books You May Enjoy

If you enjoyed this book, you may be interested in these other books by Packt:

Cloud Security Handbook

Eyal Estrin

ISBN: 9781800569195

- Secure compute, storage, and networking services in the cloud
- Get to grips with identity management in the cloud
- Audit and monitor cloud services from a security point of view
- Identify common threats and implement encryption solutions in cloud services
- Maintain security and compliance in the cloud
- Implement security in hybrid and multi-cloud environments
- Design and maintain security in a large-scale cloud environment

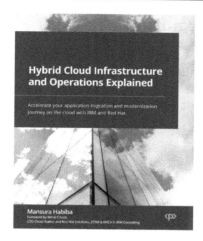

Hybrid Cloud Infrastructure and Operations Explained

Mansura Habiba

ISBN: 9781803248318

- Strategize application modernization, from the planning to the implementation phase
- Apply cloud-native development concepts, methods, and best practices
- Select the right strategy for cloud adoption and modernization
- Explore container platforms, storage, network, security, and operations
- Manage cloud operations using SREs, FinOps, and MLOps principles
- Design a modern data insight hub on the cloud

Packt is searching for authors like you

If you're interested in becoming an author for Packt, please visit `authors.packtpub.com` and apply today. We have worked with thousands of developers and tech professionals, just like you, to help them share their insight with the global tech community. You can make a general application, apply for a specific hot topic that we are recruiting an author for, or submit your own idea.

Share Your Thoughts

Now you've finished *Hybrid Cloud Security Patterns*, we'd love to hear your thoughts! Scan the QR code below to go straight to the Amazon review page for this book and share your feedback or leave a review on the site that you purchased it from.

`https://packt.link/r/1803233583`

Your review is important to us and the tech community and will help us make sure we're delivering excellent quality content.

Download a free PDF copy of this book

Thanks for purchasing this book!

Do you like to read on the go but are unable to carry your print books everywhere? Is your eBook purchase not compatible with the device of your choice?

Don't worry, now with every Packt book you get a DRM-free PDF version of that book at no cost.

Read anywhere, any place, on any device. Search, copy, and paste code from your favorite technical books directly into your application.

The perks don't stop there, you can get exclusive access to discounts, newsletters, and great free content in your inbox daily

Follow these simple steps to get the benefits:

1. Scan the QR code or visit the link below

https://packt.link/free-ebook/9781803233581

2. Submit your proof of purchase
3. That's it! We'll send your free PDF and other benefits to your email directly

www.ingramcontent.com/pod-product-compliance
Lightning Source LLC
Chambersburg PA
CBHW060540060326
40690CB00017B/3558